LiTTLE GUiDES

Space

D0354539

LiTTLE GUiDES

Space

Consultant Editor
Dr. John O'Byrne

FOG CITY PRESS

Published by Fog City Press
814 Montgomery Street
San Francisco, CA 94133 USA

Copyright © 2005 Weldon Owen Pty Ltd
First printed 2005

Chief Executive Officer: John Owen
President: Terry Newell
Publisher: Sheena Coupe
Creative Director: Sue Burk
Project Editors: Sarah Anderson, Jessica Cox, Margaret Malone
Series Design: Nika Markovtzev
Project Designers: Kathryn Morgan, Helen Perks
Editorial Coordinator: Irene Mickaiel
Index: Puddingburn Publishing Services
Production Manager: Louise Mitchell
Production Coordinator: Monique Layt
Sales Manager: Emily Jahn
Vice President International Sales: Stuart Laurence

ISBN 1 740893 48 4

Color reproduction by SC (Sang Choy) International Pte Ltd
Printed by SNP Leefung Printers Ltd
Printed in China

A Weldon Owen Production
Produced using arkiva retrieval technology
For further information, contact arkiva@weldonowen.com.au

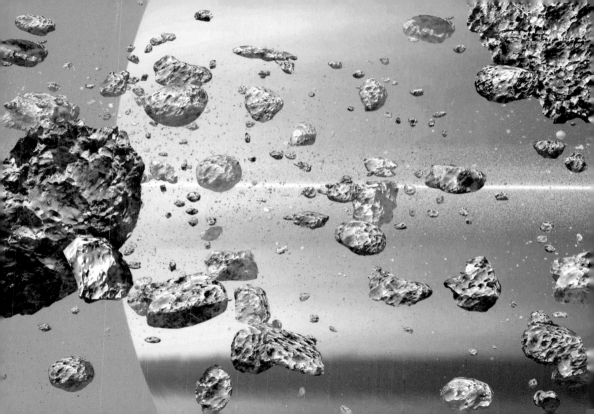

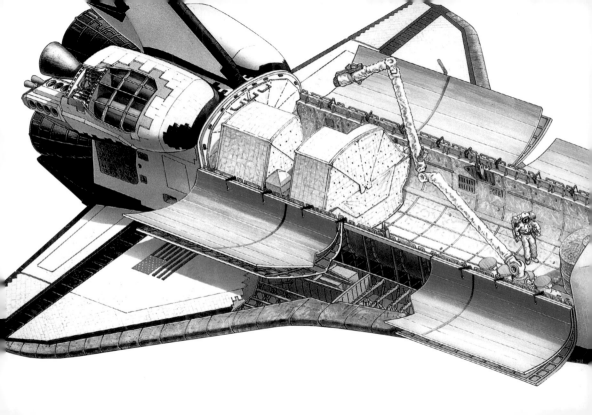

Contents

OUR SOLAR SYSTEM

Our Place in Space

In the immensity of space, our planet is a mere speck. Earth orbits the Sun, one of billions of stars in the Milky Way galaxy. The Milky Way and its companions in the Local Group are just a handful of the Universe's billions of galaxies.

1. The Local Group

EARTH'S NEIGHBORHOOD
1. Our corner of the Universe is occupied by the Local Group, a cluster of about 30 galaxies.
2. Our galaxy, the Milky Way, is a giant spiral measuring about 100,000 light-years across.
3. Earth is the third of the nine planets that orbit the Sun. The Sun itself is just one of the Milky Way's 200 billion stars.

2. The Milky Way

3. The solar system

The View from Space

From space Earth looks like a small globe suspended in the blackness of space. The closer view from our Moon reveals a beautiful blue planet. Images captured from satellites orbiting Earth show the planet in detail.

. . . AND EVEN CLOSER
A detailed image from a satellite reveals fields in a plantation in Kauai, Hawaii.

EARTH IN SPACE
The Galileo probe captured Earth and the Moon floating in space.

THEN ZOOM CLOSER . . .
From the Moon's surface, Apollo astronauts saw a blue globe.

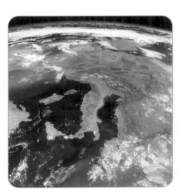

. . . AND CLOSER STILL . . .
From a space shuttle, Italy in the Mediterranean Sea is clearly visible.

The Birth of the Solar System

(1) The demise of a star probably caused a nearby cloud of gas and dust to collapse. (2) The cloud collapsed to form a dense core surrounded by a broad disk called the solar nebula, and a protosun began to form at the center. (3) Particles stuck together into fragments of rock and ice. As more of them joined, the planets gradually took shape. (4) The Sun became a star, and its radiation blew away any leftover dusty gas, leaving nine planets.

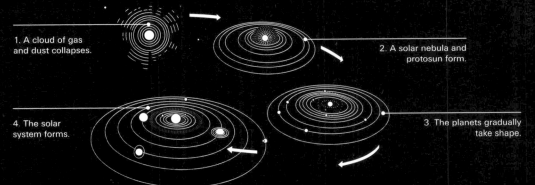

1. A cloud of gas and dust collapses.

2. A solar nebula and protosun form.

3. The planets gradually take shape.

4. The solar system forms.

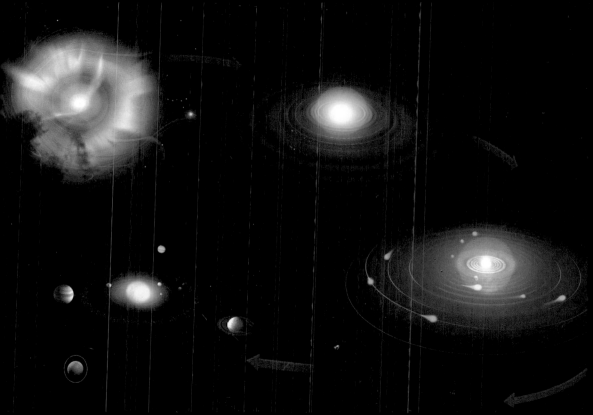

The Sun and Its Family

PLANET SIZE COMPARISON

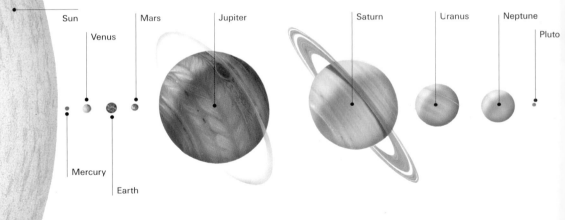

Sun

Mercury

Venus

Earth

Mars

Jupiter

Saturn

Uranus

Neptune

Pluto

THE SOLAR SYSTEM

Orbiting the Sun are the nine major planets, along with smaller bodies such as asteroids and comets. Some lie well beyond Pluto, but all are controlled by the Sun's powerful gravity. Most planets have at least one moon orbiting them. The four largest planets also have rings.

The Sun

The most important object in our sky is the Sun. Its energy powers Earth's climate and supports life. Yet, the Sun is an ordinary star like a billion others in the Milky Way galaxy.

CROSS SECTION

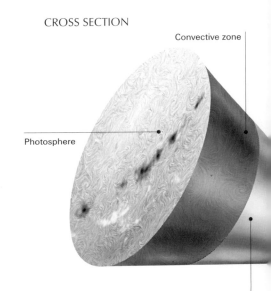

Photosphere

Convective zone

Radiative zone

SIZE COMPARISON

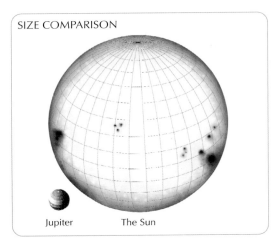

Jupiter

The Sun

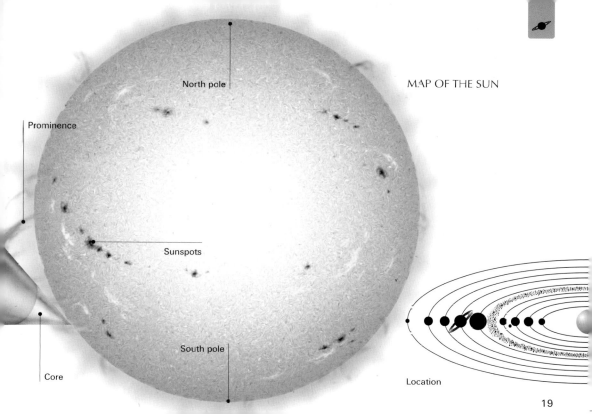

North pole

Prominence

MAP OF THE SUN

Sunspots

South pole

Core

Location

19

Solar Activity

It is not easy to see activity on the surface of the Sun with the naked eye. Special filters enable us to see sunspots, prominences, and solar flares.

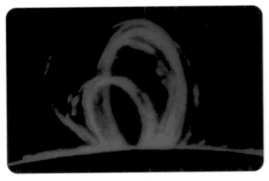

LOOPS OF GAS
At times the Sun's powerful magnetic field holds gas loops called prominences high above the surface of the Sun.

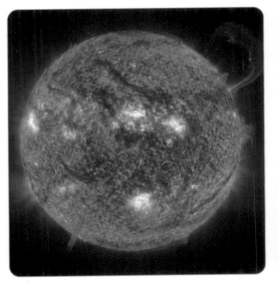

THE SUN'S ATMOSPHERE
The corona is the Sun's atmosphere. Sometimes, prominences or solar flares erupt through the corona.

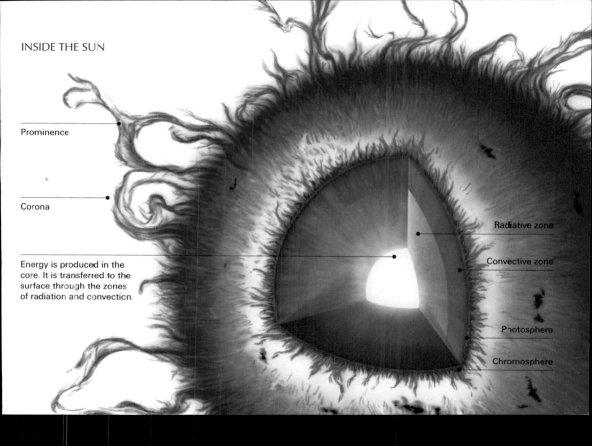

INSIDE THE SUN

Prominence

Corona

Energy is produced in the core. It is transferred to the surface through the zones of radiation and convection.

Radiative zone

Convective zone

Photosphere

Chromosphere

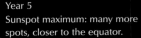

Sunspots

Sunspots are dark because they are very much cooler than the surrounding surface. They form where high magnetic force stops heat from reaching the surface. The number of sunspots waxes and wanes every 11 years. As the last spots of one cycle fade, the first of a new cycle appear.

SUNSPOT CYCLE

Year 1
Sunspot minimum: few spots at high latitudes.

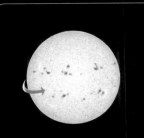

Year 5
Sunspot maximum: many more spots, closer to the equator.

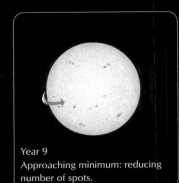

Year 9
Approaching minimum: reducing number of spots.

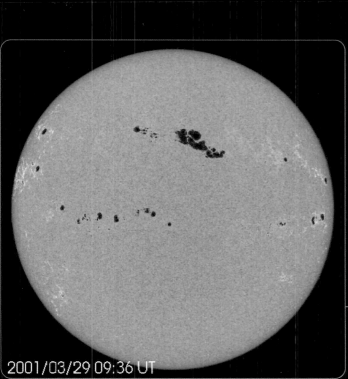

2001/03/29 09:36 UT

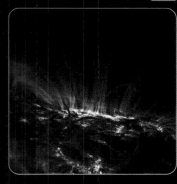

HOT SPOTS
The bright glowing gas flowing in the corona above sunspots has a temperature of almost 2 million degrees Fahrenheit (1 million degrees Celsius).

GIANT SUNSPOTS
The sunspot area visible at top center in this image is 13 times larger than the entire surface of Earth! It was the source of many solar flares.

Mercury

Mercury is the closest planet to the Sun. Mercury's sunny side is very hot—temperatures can reach 800°F (427°C), but since there is little atmosphere to trap the Sun's energy, they plunge to -280°F (-173°C) on its night side.

SIZE COMPARISON

Mercury   Earth

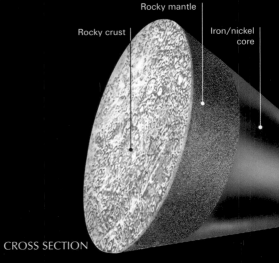

Rocky mantle

Rocky crust

Iron/nickel core

CROSS SECTION

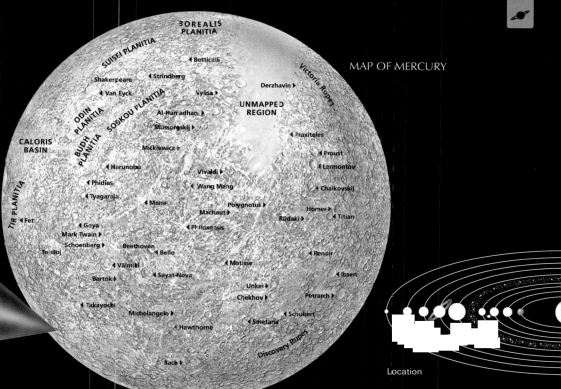

MAP OF MERCURY

BOREALIS PLANITIA

SUISEI PLANITIA

Botticelli

Strindberg

Shakespeare

Van Eyck

Vyāsa

Derzhavin

Victoria Rupes

UNMAPPED REGION

ODIN PLANITIA

SOBKOU PLANITIA

Al-Hamadhan

Mussorgskij

BUDH PLANITIA

CALORIS BASIN

Mickiewicz

Praxiteles

Harunobu

Phidias

Tyagaraja

Vivaldi

Wang Meng

Mena

Polygnotus

Machaut

Proust

Lermontov

Chaikovskij

Homer

Rūdaki

Titian

TIR PLANITIA

Fet

Goya

Mark Twain

Schoenberg

Tolstoj

Beethoven

Bello

Philoxenus

Vālmiki

Matisse

Renoir

Ibsen

Bartók

Sayat-Nova

Unkei

Chekhov

Petrarch

Takayoshi

Michelangelo

Smetana

Schubert

Hawthorne

Bach

Discovery Rupes

Location

25

Mercury's Surface

In 1974–75, photos from the Mariner 10 spacecraft showed that Mercury's surface looks like the Moon. It is covered with craters and basins—the scars of impacts. The biggest basin, Caloris, is 800 miles (1,300 km) across.

After the planet formed and cooled, its core shrank.

The crust buckled, pushing up giant wrinkles of rock called scarps.

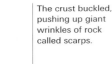

MERCURY'S WRINKLES

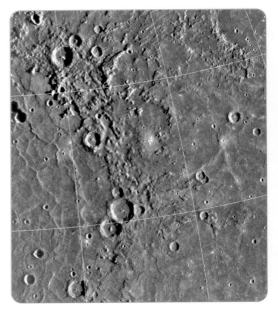

IMPACT CRATERS
This view of the surface of Mercury shows some of the thousands of impact craters that cover the planet.

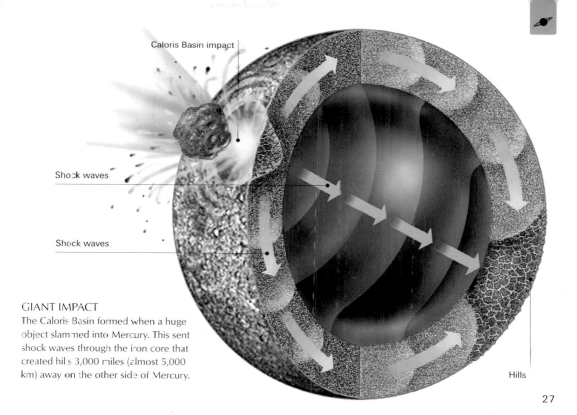

Caloris Basin impact

Shock waves

Shock waves

GIANT IMPACT

The Caloris Basin formed when a huge
object slammed into Mercury. This sent
shock waves through the iron core that
created hills 3,000 miles (almost 5,000
km) away on the other side of Mercury.

Hills

Mercury's Path

Mercury's year—one orbit of the Sun—is only 88 Earth days long, but its day—one rotation on its axis—lasts 59 Earth days. A full solar day from noon to noon lasts 176 Earth days.

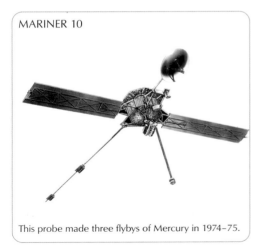

MARINER 10

This probe made three flybys of Mercury in 1974–75.

MERCURY IN TRANSIT
Occasionally, as seen from Earth, Mercury's orbit carries it in front of the Sun—an event known as a transit.

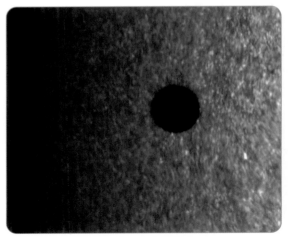

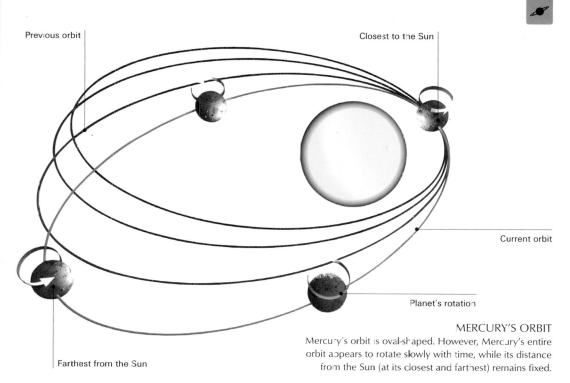

Previous orbit

Closest to the Sun

Current orbit

Planet's rotation

Farthest from the Sun

MERCURY'S ORBIT

Mercury's orbit is oval-shaped. However, Mercury's entire orbit appears to rotate slowly with time, while its distance from the Sun (at its closest and farthest) remains fixed.

29

Venus

Venus is the second planet from the Sun, and the one most like Earth in size. On its surface are mountain ranges, volcanoes, and lava flows. Despite being farther from the Sun than Mercury, Venus is hotter.

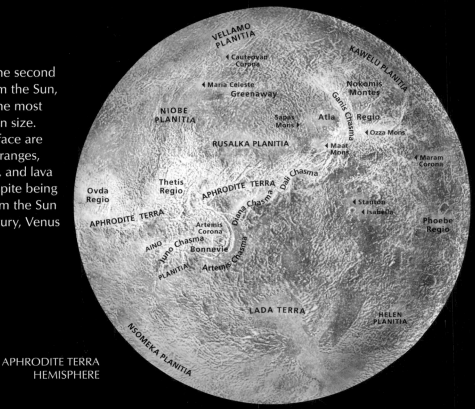

VELLAMO PLANITIA

KAWELU PLANITIA

◄ Cauteovan Corona

◄ Maria Celeste Greenaway

Nokomis Montes

NIOBE PLANITIA

Sapas Mons ►

Atla Regio

Ganis Chasma

◄ Ozza Mons

RUSALKA PLANITIA

◄ Maat Mons

◄ Maram Corona

Ovda Regio

Thetis Regio

APHRODITE TERRA

Dali Chasma

APHRODITE TERRA

Diana Chasma

◄ Stanton
◄ Isabella

Phoebe Regio

Artemis Corona

AINO

Juno Chasma

Bonnevie

Artemis Chasma

PLANITIA

LADA TERRA

HELEN PLANITIA

NSOMEKA PLANITIA

APHRODITE TERRA
HEMISPHERE

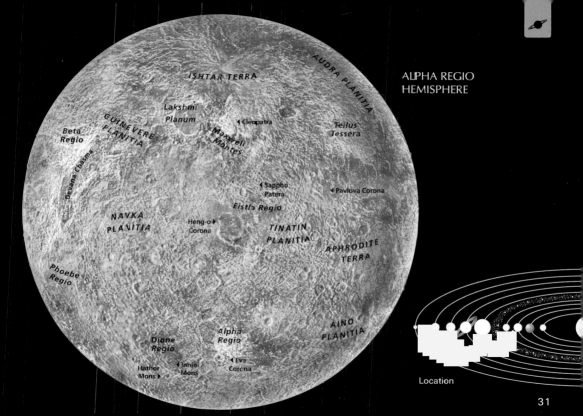

ALPHA REGIO
HEMISPHERE

AUDRA PLANITIA

ISHTAR TERRA

Lakshmi
Planum

◄ Cleopatra

Maxwell
Montes

Tellus
Tessera

GUINEVERE
PLANITIA

Beta
Regio

Devana Chasma

◄ Sappho
Patera

◄ Pavlova Corona

Eistla Regio

NAVKA
PLANITIA

Heng-o ►
Corona

TINATIN
PLANITIA

APHRODITE
TERRA

Phoebe
Regio

AINO
FLANITIA

Dione
Regio

Alpha
Regio

Hathor
Mons ►

◄ Innini
Mons

◄ Eve
Corona

Location

31

Venus' Features

We often see Venus looking like a bright star in the morning or evening sky. The brightness is from sunlight reflected off a layer of white sulfuric acid clouds above the planet's surface.

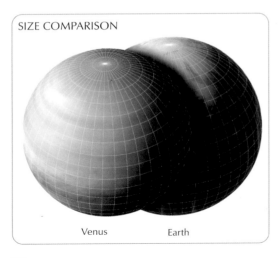

SIZE COMPARISON

Venus   Earth

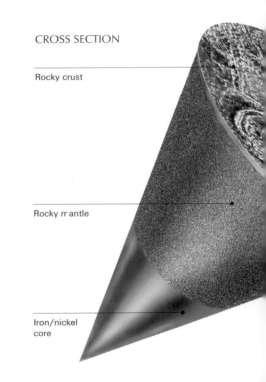

CROSS SECTION

Rocky crust

Rocky mantle

Iron/nickel core

GREENHOUSE EFFECT

Venus suffers from a greenhouse effect. Strong sunlight filters through the clouds and heats the surface, but the clouds and carbon dioxide in the atmosphere keep the heat from escaping back into space. Venus cannot cool down.

Clouds reflect much of the Sun's energy.

Some solar energy passes through clouds and heats the surface.

Carbon dioxide keeps heat from escaping.

Mapping Venus

Between 1978 and 1994, the Pioneer, Venera, and Magellan spacecraft used radar to map Venus's surface.

MAAT MONS
This computer image shows Maat Mons, a 5-mile (8-km) high volcano.

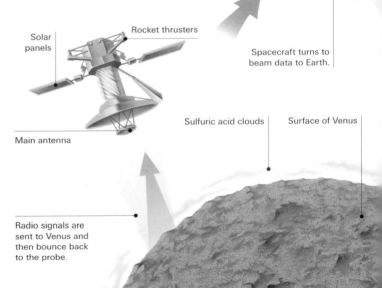

Solar panels

Rocket thrusters

Spacecraft turns to beam data to Earth.

Main antenna

Sulfuric acid clouds

Surface of Venus

Radio signals are sent to Venus and then bounce back to the probe.

Data are received
on Earth.

Deep-space
tracking antenna

Data are sent to
mission control.

MAGELLAN PROBE

NASA's Magellan probe beamed radar
signals at Venus. The radio waves hit
the surface, and bounced back to the
probe. Using the reflected radio
waves, the probe helped create a
picture of the surface of Venus.

Scientist adds new
data to map.

Completed map
of Venus

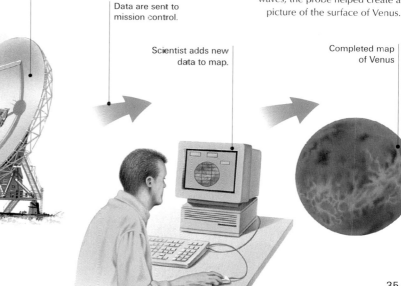

Earth

The third planet from the Sun, Earth is the largest of the small rocky planets. It could be called the water planet. Oceans cover 71 percent of Earth's surface. From space, the oceans can be seen amid the white of water ice and swirling clouds.

EASTERN HEMISPHERE

EUROPE
ASIA
Siberia
Plateau of Tibet
Hindu Kush
Honshu
Yangtze Basin
Arabian Peninsula
Arabian Sea
Deccan Plateau
Bay of Bengal
Mekong Basin
South China Sea
PACIFIC OCEAN
Philippine Islands
AFRICA
Sri Lanka
Borneo
New Guinea
Sumatra
Java Trench
Java
INDIAN OCEAN
Madagascar
AUSTRALIA
SOUTHERN OCEAN
ANTARCTICA

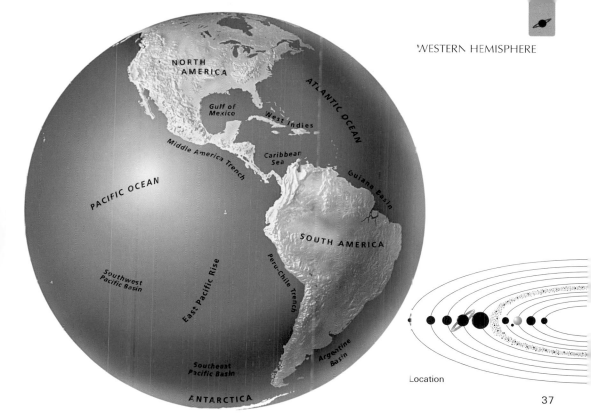

NORTH AMERICA

Gulf of Mexico

West Indies

ATLANTIC OCEAN

Middle America Trench

Caribbean Sea

Guiana Basin

PACIFIC OCEAN

SOUTH AMERICA

Southwest Pacific Basin

East Pacific Rise

Peru-Chile Trench

Southeast Pacific Basin

Argentine Basin

ANTARCTICA

Location

Earth's Structure

Earth's stiff crust is broken into pieces called plates. Thin plates lie under the oceans, and thick plates carry continents. Heat from the core drives mantle rocks in a slow boiling motion that pushes the plates together and pulls them apart.

When two plates with ocean crust collide, magma breaks through the crust, forming lines of volcanoes called island arcs.

As ocean plates are pushed apart, magma rises through the gap and cools and hardens to form a mid-ocean ridge.

Sometimes magma from a spot deep in the mantle bursts through the middle of a plate, forming a volcano known as a hot-spot volcano.

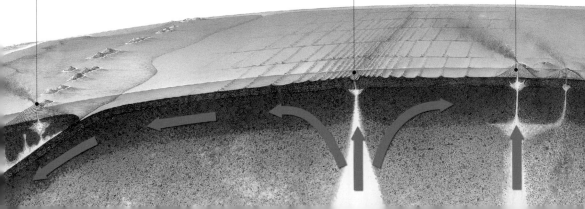

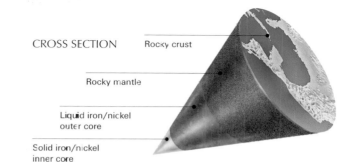

CROSS SECTION

Rocky crust

Rocky mantle

Liquid iron/nickel
outer core

Solid iron/nickel
inner core

When ocean crust
meets continental crust,
the thin ocean crust is
pushed down into the
mantle. Volcanoes may
form above the
downgoing crust.

If two plates
carrying continental
crust collide, the
land buckles and
folds, creating
mountains.

The Story of Earth

4.6–4.2 BILLION YEARS AGO

4.2–3.8 BILLION YEARS AGO

200 MILLION YEARS AGO

The young Earth's surface is bombarded by comets and meteorites. The hot young planet cools and the heavier elements sink to the core.

Bombardment slows. Lava flows released by the impacts cool and form the crust. Water from volcanoes and comets slowly forms the oceans.

The landmasses of the Northern and Southern hemispheres are joined in the single supercontinent of Pangaea, surrounded by one large ocean.

90 MILLION YEARS AGO

TODAY

50 MILLION YEARS FROM NOW

Pangaea has broken up and the South Atlantic Ocean has opened, but North America remains close to Europe, and Australia remains joined to Antarctica.

The continents reached their present positions by roughly 12 million years ago. Continental collisions will eventually result in a new supercontinent.

The Atlantic widens, the Mediterranean vanishes as Africa and Europe join up, Australia and Southeast Asia collide, and California slides up to Alaska.

Earth and the Sun

As Earth spins on its axis, we get day and night, and as it moves around the Sun, we get changing seasons.

DAY AND NIGHT
Day begins when our part of the world turns to face the Sun, and night falls when it turns away.

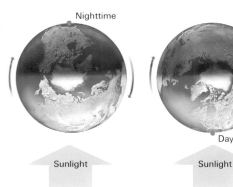

Nighttime

Daytime

Sunlight

Sunlight

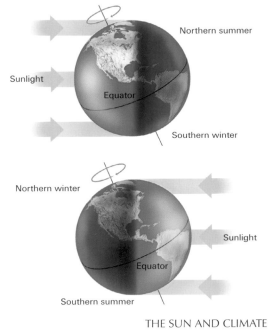

Northern summer

Sunlight

Equator

Southern winter

Northern winter

Sunlight

Equator

Southern summer

THE SUN AND CLIMATE
At the equator, the Sun's rays hit Earth almost directly. Farther away, the rays hit at an angle and the climate is cooler.

THE SEASONS

The seasons change as the amount of sunlight falling on each part of Earth changes.

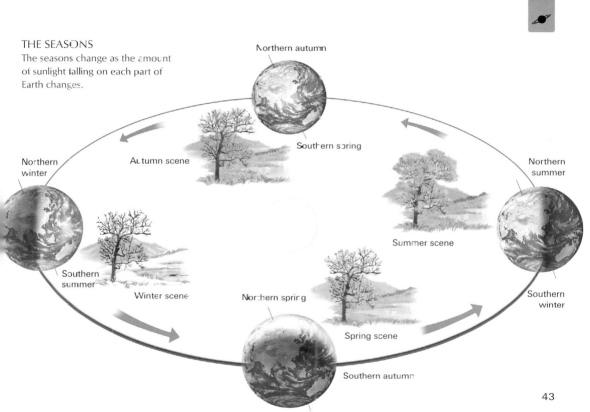

Northern autumn

Southern spring

Autumn scene

Northern winter

Southern summer

Winter scene

Northern spring

Southern autumn

Spring scene

Northern summer

Southern winter

Summer scene

Earth's Atmosphere

Earth's atmosphere is mostly nitrogen and oxygen with traces of water vapor. Earth is the only planet in the solar system with an atmosphere that contains sufficient water and oxygen for life to flourish.

WHERE SPACE BEGINS
No sharp line separates space from Earth's atmosphere—the air just gradually becomes thinner with altitude. The atmosphere's last traces can be detected roughly 600 miles (1,000 km) above the ground.

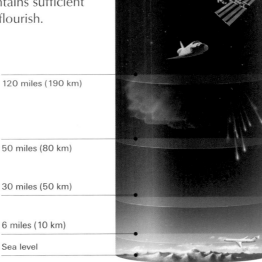

Exosphere

120 miles (190 km)

Thermosphere

50 miles (80 km)

Mesosphere

30 miles (50 km)

Stratosphere

6 miles (10 km)

Sea level

Troposphere

MAGNETIC FIELD

Generated by the iron in Earth's core, a magnetic field surrounds our planet like a shield. It helps to protect Earth from the charged particles of the solar wind.

WEATHER

This swirling cloud is low in the atmosphere. As Earth's land and water absorb heat from the Sun, they control our weather.

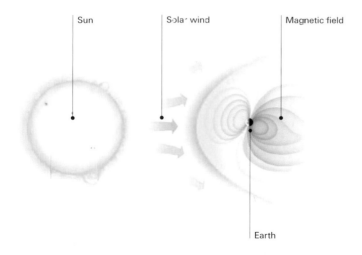

Sun | Solar wind | Magnetic field

Earth

Our Moon

The Moon is Earth's only natural satellite. The nearside hemisphere is the side of the Moon that always faces Earth. The farside hemisphere was revealed only when probes first visited our Moon.

NEARSIDE HEMISPHERE

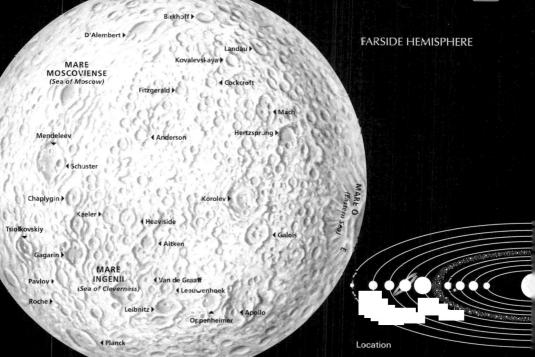

Birkhoff ▶

D'Alembert ▶

Landau ▶

Kovalevskaya ▶

MARE
MOSCOVIENSE
(Sea of Moscow)

◀ Cockcroft

Fitzgerald ▶

◀ Mach

Hertzsprung ▶

Mendeleev ▾

◀ Anderson

◀ Schuster

Chaplygin ▶

Korolev ▶

Kaeler ▶

Tsiolkovskiy ▾

◀ Heaviside

◀ Galois

Gagarin ▶

◀ Aitken

MARE
INGENII
(Sea of Cleverness)

◀ Van de Graaff

Pavlov ▶

◀ Leeuwenhoek

Roche ▶

Leibnitz ▶

◀ Apollo

Oppenheimer ▾

◀ Planck

MARE O
(Eastern Sea)
E

Location

Our Moon's Features

The Moon is covered with craters and basins caused by meteorite impacts. Dark lava (molten rock) flowed from under the surface to fill many of these basins. The youngest craters have rays, streaks of rocks splashed across the lava surfaces.

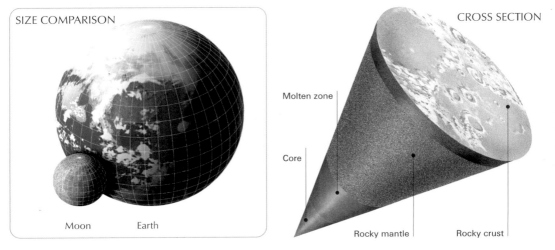

SIZE COMPARISON

Moon　　　Earth

CROSS SECTION

Molten zone

Core

Rocky mantle

Rocky crust

HOW THE MOON FORMED

The Moon formed after Earth did, from the debris that flew out when another body slammed into Earth. The debris circled around Earth, then clumped together to form the Moon.

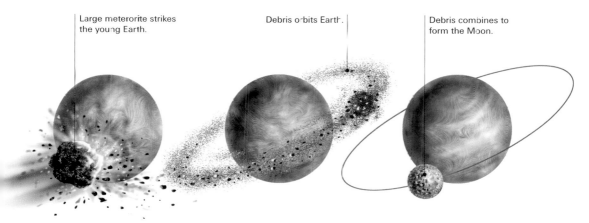

Large meteorite strikes the young Earth.

Debris orbits Earth.

Debris combines to form the Moon.

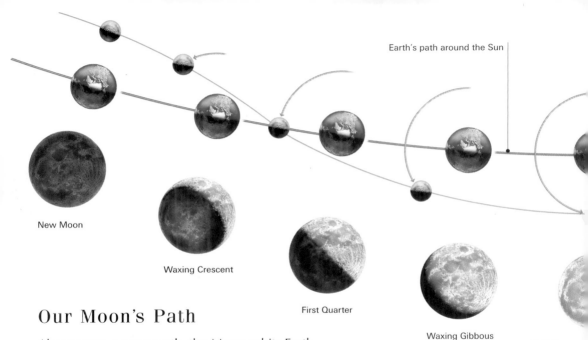

Earth's path around the Sun

New Moon

Waxing Crescent

First Quarter

Waxing Gibbous

Full Moon

Our Moon's Path

About once every month, the Moon orbits Earth.
It always keeps the same side turned toward us, although
the amount of the surface lit by the Sun changes.

50

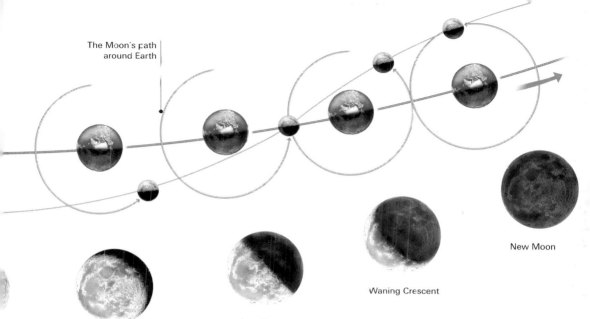

The Moon's path around Earth

New Moon

Waning Crescent

Last Quarter

Waning Gibbous

PHASES OF THE MOON

Between New and Full Moon, the sunlit part of the Moon grows larger (waxing). Between the Full Moon and the next New Moon, it grows smaller (waning).

Solar Eclipses

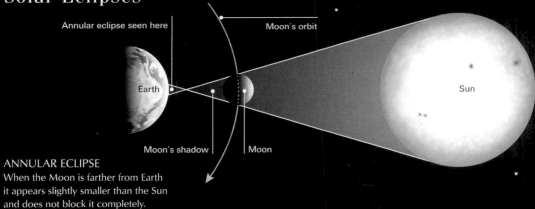

Annular eclipse seen here

Moon's orbit

Earth

Sun

Moon's shadow

Moon

ANNULAR ECLIPSE
When the Moon is farther from Earth it appears slightly smaller than the Sun and does not block it completely.

Total eclipse seen here

Moon's orbit

Earth

Sun

Moon's shadow

Moon

TOTAL ECLIPSE

During a total eclipse, the Moon passes directly
between Earth and the Sun, and its shadow passes
over part of Earth. For people in the narrow path
of the shadow, the Sun is blacked out totally.

DEVELOPMENT OF AN ECLIPSE

The Moon passes in front of the Sun until it blocks
out the disk of the Sun, revealing the corona.

Lunar Eclipses

When the Moon passes directly behind Earth, Earth blocks direct sunlight. The Moon darkens but doesn't disappear altogether—instead, it glows a deep copper-orange, colored by sunlight that filters through Earth's atmosphere.

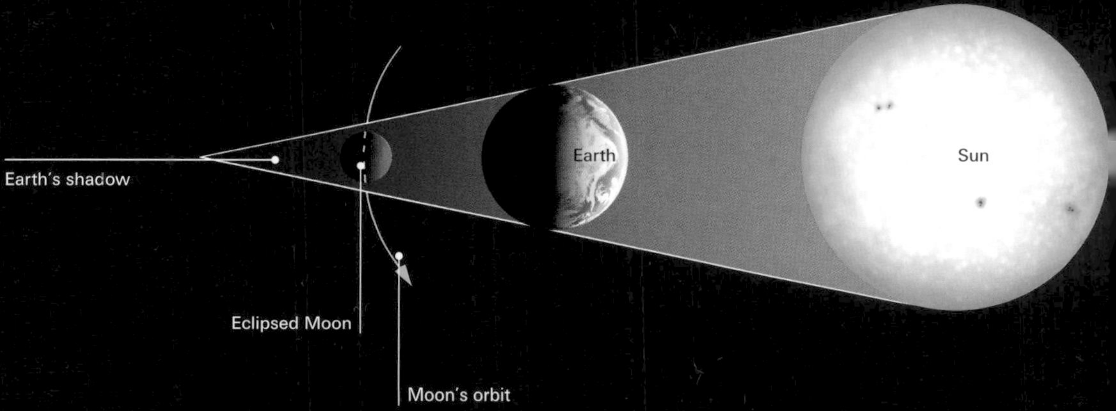

Earth's shadow

Earth

Sun

Eclipsed Moon

Moon's orbit

TOTAL LUNAR ECLIPSE

STAGES OF AN ECLIPSE
This time-lapse photo reveals the stages in a total lunar eclipse, when the Moon passes fully into the darkest part of Earth's shadow.

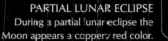

PARTIAL LUNAR ECLIPSE
During a partial lunar eclipse the Moon appears a coppery red color.

Meteors and Meteorites

More than 200 tons (180 tonnes) of space debris enter Earth's atmosphere every day. Most of the material burns up in the atmosphere. The result is a bright streak of light called a meteor, or shooting star. Before it enters the atmosphere, a piece of debris is called a meteoroid. If it survives to hit the ground, the debris is called a meteorite.

IMPACT CRATERS
Most meteors burn up in the atmosphere, but big ones survive to reach the ground. Every few thousand years, a really big object may produce an impact crater.

Stony meteorite

Mars meteorite ALH 84001

Iron meteorite

Mars

Mars is the planet most like our own. It has four seasons, polar ice caps, channels carved by water, and a rotation time just 41 minutes longer than Earth's.

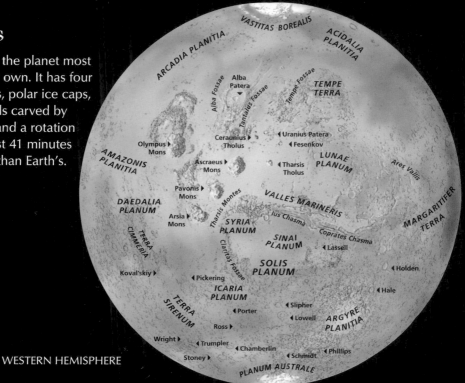

VASTITAS BOREALIS

ACIDALIA PLANITIA

ARCADIA PLANITIA

TEMPE TERRA

Alba Fossae
Alba Patera
Tantalus Fossae
Tempe Fossae

Ceraunius Tholus

Uranius Patera
Fesenkov

Olympus Mons

AMAZONIS PLANITIA

Ascraeus Mons

Tharsis Tholus

LUNAE PLANUM

Ares Vallis

Pavonis Mons

DAEDALIA PLANUM

Tharsis Montes

VALLES MARINERIS

MARGARITIFER TERRA

Arsia Mons

Ius Chasma

SYRIA PLANUM

Coprates Chasma

TERRA CIMMERIA

SINAI PLANUM

Lassell

Claritas Fossae

SOLIS PLANUM

Koval'skiy

Pickering

ICARIA PLANUM

Holden

Hale

Porter

Slipher

TERRA SIRENUM

Lowell

ARGYRE PLANITIA

Ross

Wright
Trumpler
Chamberlin
Schmidt
Phillips

Stoney

PLANUM AUSTRALE

WESTERN HEMISPHERE

58

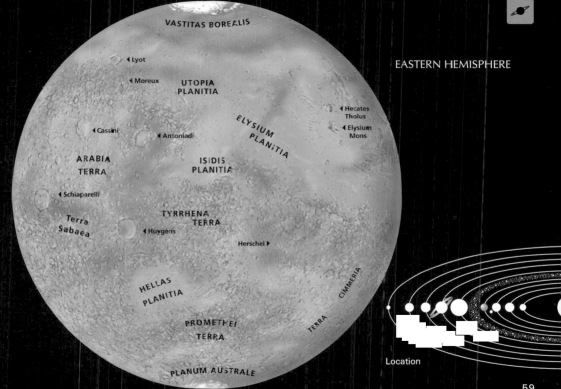

VASTITAS BOREALIS

◄ Lyot

◄ Moreux   UTOPIA
                PLANITIA

                              ELYSIUM
                              PLANITIA
◄ Cassini   ◄ Antoniadi

ARABIA   ISIDIS
TERRA   PLANITIA

◄ Schiaparelli

Terra   TYRRHENA
Sabaea   TERRA
                ◄ Huygens

                        Herschel ►

HELLAS
PLANITIA

PROMETHEI   TERRA   CIMMERIA
TERRA

PLANUM AUSTRALE

EASTERN HEMISPHERE

◄ Hecates
  Tholus

◄ Elysium
  Mons

Location

Mars' Features

Mars looks red in the sky, earning it the nickname the Red Planet. The color comes from its rusty-orange rocks and fine red sand. It is a cold desert—temperatures can reach 81°F (27°C) by day but drop to –190°F (–123°C) at night.

SIZE COMPARISON

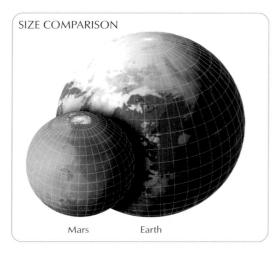

Mars   Earth

Rocky crust

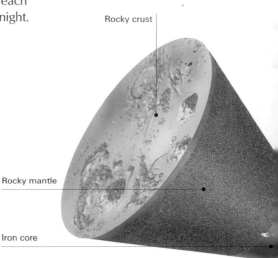

Rocky mantle

Iron core

CROSS SECTION

MARTIAN SURFACE

The southern half of Mars is heavily cratered. The craters below lie within the Hellas Planitia region. This ancient impact basin is the largest basin on Mars. It was formed when a large asteroid or comet hit the surface.

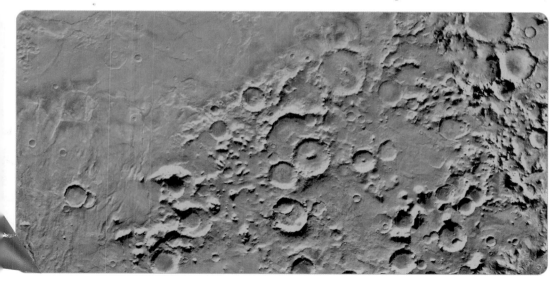

The Moons of Mars

Phobos and Deimos, the two Martian moons, are covered in craters and a layer of dusty shattered rock. Phobos is so close to Mars that it completes more than three orbits each Martian day.

THE SURFACE OF PHOBOS
The larger of Mars' two satellites, Phobos is covered with craters, the largest of which is called Stickey.

MARTIAN MOONS

Deimos

Phobos

Phobos and Deimos have irregular shapes. Phobos is 16 miles (26 km) across, while Deimos is about half that size.

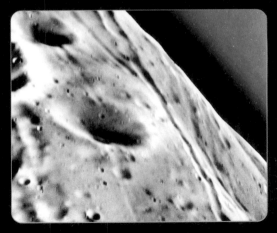

MARTIAN SATELLITES IN ORBIT

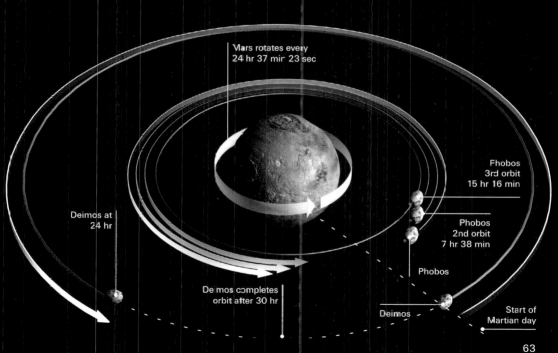

Mars rotates every
24 hr 37 min 23 sec

Phobos
3rd orbit
15 hr 16 min

Phobos
2nd orbit
7 hr 38 min

Phobos

Deimos at
24 hr

Deimos completes
orbit after 30 hr

Deimos

Start of
Martian day

Asteroids

Between Mars and Jupiter, there are millions of asteroids—small bodies made of rock or metal. Not all asteroids orbit in this main belt. The Trojan asteroids travel in Jupiter's orbit. Other asteroids have wide elliptical orbits that cross the orbit of Earth.

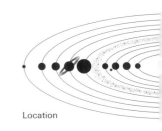

Location

ASTEROID BELT
The strong gravity of Jupiter may have kept asteroids in the main belt from clumping together to become a planet like Earth or Mars.

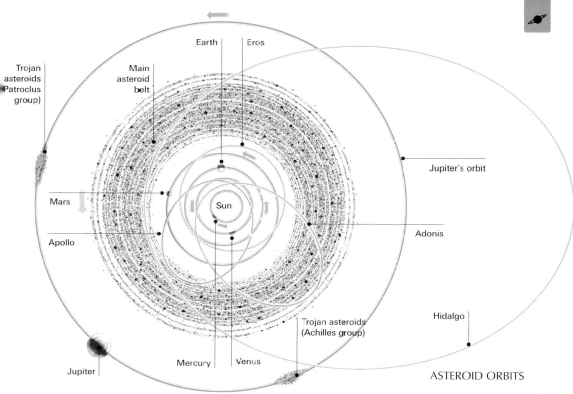

Trojan asteroids (Patroclus group)

Earth

Eros

Main asteroid belt

Jupiter's orbit

Mars

Sun

Adonis

Apollo

Hidalgo

Trojan asteroids (Achilles group)

Jupiter

Mercury

Venus

ASTEROID ORBITS

65

Jupiter

The most massive of all the planets, Jupiter is an enormous ball of gas—mostly hydrogen and helium, like the Sun, along with small amounts of water, methane, and ammonia. This giant planet lacks a solid surface.

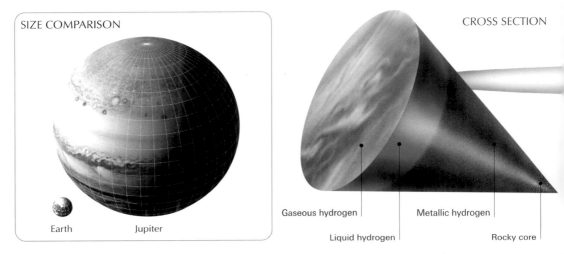

SIZE COMPARISON

Earth   Jupiter

CROSS SECTION

Gaseous hydrogen

Liquid hydrogen

Metallic hydrogen

Rocky core

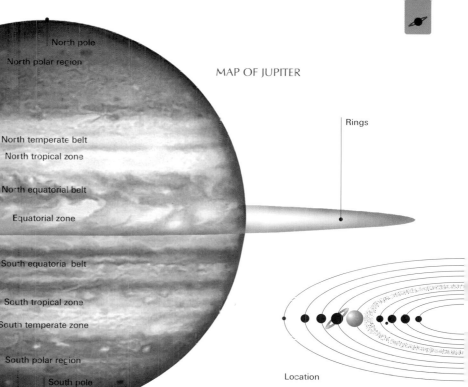

North pole
North polar region

MAP OF JUPITER

North temperate belt
North tropical zone
North equatorial belt
Equatorial zone

Great Red Spot

South equatorial belt
South tropical zone
South temperate zone
South polar region
South pole

Rings

Location

67

Jupiter's Features

Jupiter spins rapidly in space. This creates strong winds that whip its clouds into bands.

JUPITER'S BANDS

The light bands are called zones. The dark bands, known as belts, show deeper layers.

The Great Red Spot is a giant storm that thrives on the energy of Jupiter's winds and its core. It has been visible for at least 300 years.

THE RINGS OF JUPITER

Jupiter's rings are very thin, and some cannot be seen from Earth. There are three parts—the halo, the main ring, and the Gossamer ring.

Halo   Main   Gossamer

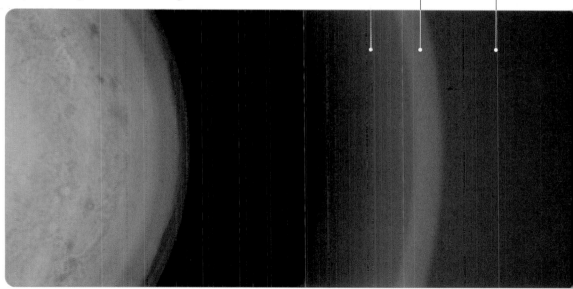

The Moons of Jupiter

In 1610, astronomer Galileo Galilei discovered Jupiter's
four largest moons—Io, Europa, Ganymede, and Callisto.
Today, we know of 61 moons, of which only 34 have names.
Most of the moons are asteroids captured by the pull of
Jupiter's gravity.

CALLISTO

EUROPA

GANYMEDE

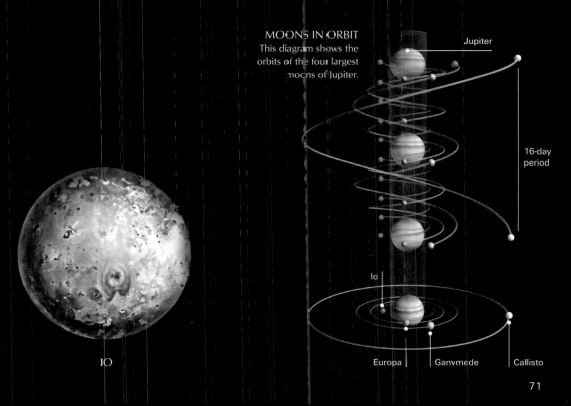

MOONS IN ORBIT
This diagram shows the orbits of the four largest moons of Jupiter.

Jupiter

16-day period

Io

IO

Europa   Ganymede   Callisto

71

Saturn

Saturn's three obvious rings are made up of hundreds
of individual ringlets. Near the surface, Saturn's hydrogen
and helium are gaseous, but they become fluid deeper in.
The rocky core is about twice as hot as the Sun's surface.

Rocky core

Metallic hydrogen

Liquid hydrogen

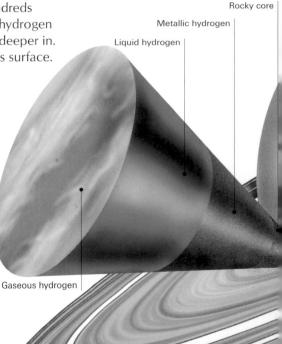

SIZE COMPARISON

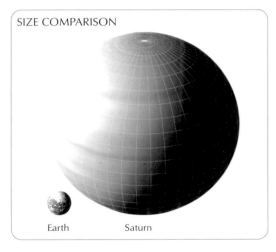

Earth   Saturn

Gaseous hydrogen

72

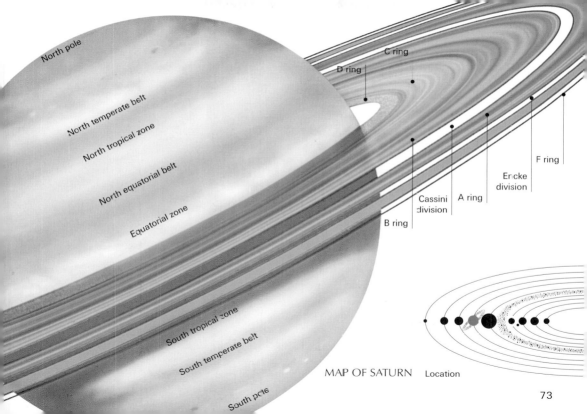

North pole

North temperate belt

North tropical zone

North equatorial belt

Equatorial zone

South tropical zone

South temperate belt

South pole

C ring

B ring

B ring

Cassini division

A ring

Encke division

F ring

MAP OF SATURN   Location

73

RINGS AND RINGLETS
The many dark and bright ringlets are the result of different sizes and brightness of the icy fragments.

HOW THE RINGS FORMED
One theory suggests the rings are only a few million years old and were formed by a cosmic collision.

1. A large object smashes into an icy moon orbiting Saturn.

Saturn's Rings

Saturn is known as the Ringed Planet. Jupiter, Uranus, and Neptune also have rings, but Saturn's are the most magnificent. Saturn's rings are made of chunks of ice and rock. They are thousands of miles wide but only 40 to 400 feet (12 to 120 m) thick.

2. The impact shatters the moon and the particles then orbit Saturn.

3. Collisions among the particles grind them into smaller pieces.

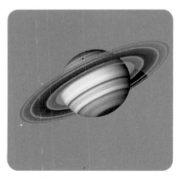

4. Over the years, these particles spread out to form the broad rings.

The Moons of Saturn

Saturn has at least 31 moons, ranging in size from mere rocks a few miles across to Titan, the second-largest moon in the solar system. Like Jupiter's moons, some of Saturn's moons are asteroids captured by the pull of Saturn's gravity.

ENCELADUS

IAPETUS

MIMAS

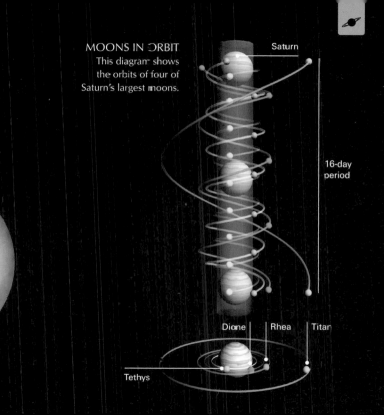

MOONS IN ORBIT
This diagram shows
the orbits of four of
Saturn's largest moons.

Saturn

16-day
period

Dione

Rhea

Titan

Tethys

TITAN

Uranus

So far, the face of Uranus has appeared mostly featureless, but that is changing as we learn more about the planet. Beneath Uranus' gaseous top layer is a dense liquid layer and a hot rocky core. The planet is circled by 11 rings.

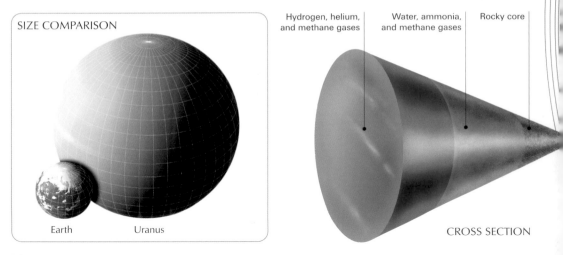

SIZE COMPARISON

Earth   Uranus

Hydrogen, helium, and methane gases

Water, ammonia, and methane gases

Rocky core

CROSS SECTION

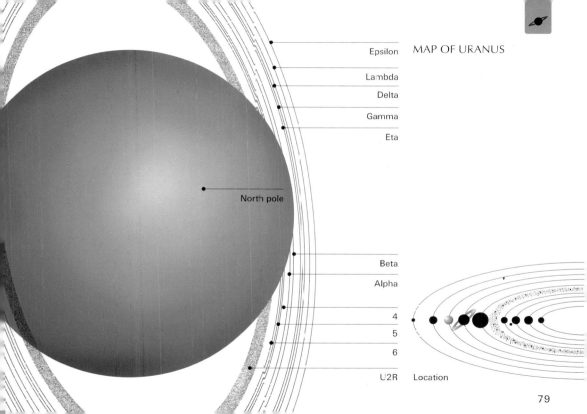

MAP OF URANUS

Epsilon

Lambda

Delta

Gamma

Eta

North pole

Beta

Alpha

4

5

6

U2R   Location

Uranus' Features

Uranus might be called the planet-on-its-side—its axis is
tipped 98 degrees. Uranus' orbit takes 84 Earth years,
so each pole receives a long period of constant sunlight
followed by a long period of darkness.

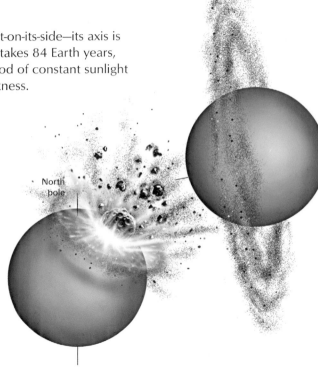

North
pole

North
pole

THE SIDEWAYS PLANET
Scientists speculate that as Uranus
was forming, a large object struck and
knocked it over on its side. The impact
may also have created the moons and
rings that orbit Uranus.

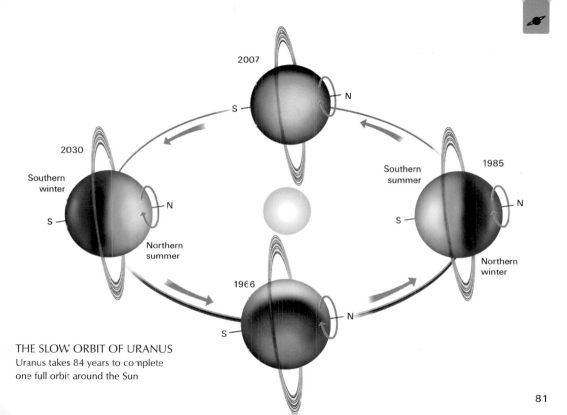

2007

S   N

2030
Southern
winter

S   N

Northern
summer

1985
Southern
summer

S   N

Northern
winter

1966

S   N

THE SLOW ORBIT OF URANUS
Uranus takes 84 years to complete
one full orbit around the Sun

The Moons of Uranus

Uranus has a total of 27 known moons. The largest five
are Miranda, Ariel, Umbriel, Titania, and Oberon.
All except these five large moons measure just a few tens
of miles across, and are likely to be captured asteroids.

ARIEL

MIRANDA

UMBRIEL

TITANIA

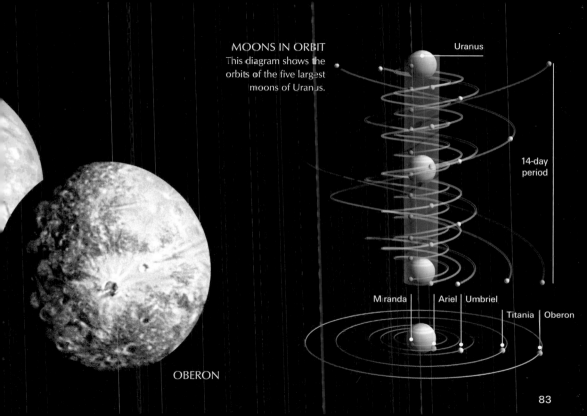

MOONS IN ORBIT
This diagram shows the orbits of the five largest moons of Uranus.

Uranus

14-day period

Miranda | Ariel | Umbriel

Titania | Oberon

OBERON

83

Neptune

Neptune is colored blue by traces of methane in its gaseous atmosphere. Beneath the clouds, the planet probably has a deep "ocean" of water, ammonia, and methane and a hot rocky core. Neptune is circled by five faint rings.

SIZE COMPARISON

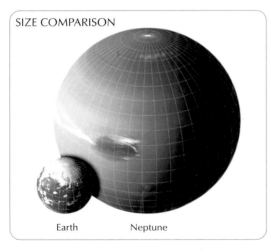

Earth   Neptune

CROSS SECTION

Water, ammonia, and methane slush

Hydrogen, helium, and methane gases

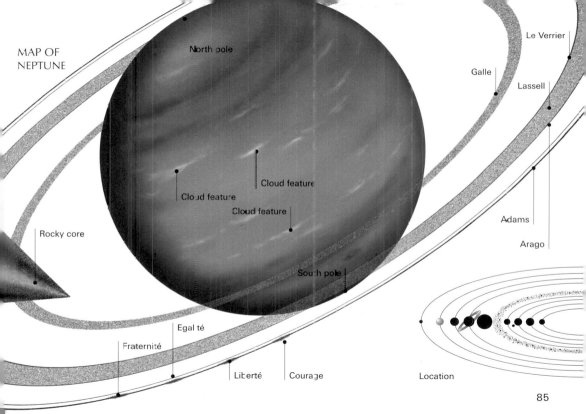

MAP OF
NEPTUNE

North pole

Le Verrier

Galle

Lassell

Cloud feature

Cloud feature

Cloud feature

Rocky core

Adams

Arago

South pole

Egalité

Fraternité

Liberté

Courage

Location

85

Neptune's Features

Neptune is a ball of hydrogen, helium, and methane. The tilt of Neptune's axis (29.6 degrees) is not as extreme as Uranus's, so its seasonal changes are less dramatic. Neptune has raging storms and clouds.

NEPTUNE'S RINGS
The rings are made up of particles that range from microscopic to the size of a house. The gravity of Neptune's moons keeps the rings in place.

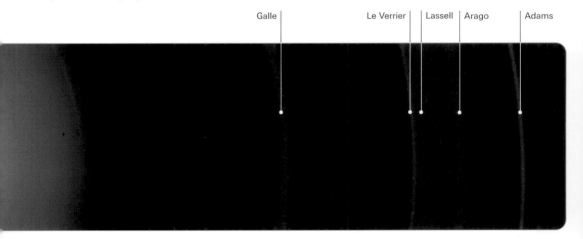

Galle | Le Verrier | Lassell | Arago | Adams

GREAT DARK SPOT

This 1989 photo shows the giant storm in Neptune's atmosphere known as the Great Dark Spot (center) with its companion, the White Scooter. These features have vanished since the photo was taken.

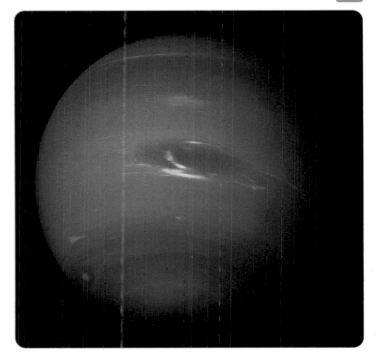

DID YOU KNOW?
Since Neptune was first spotted in 1846, it still has not made a full orbit around the Sun. The trip takes 165 Earth years.

The Moons of Neptune

Not many of Neptune's moons are known because the planet is so far from the Sun that it is hard to see them. We know of 13 moons, of which five have yet to be named. Triton is the only large satellite.

NEREID

TRITON

MOONS IN ORBIT
This illustration shows the orbits of three of Neptune's moons—Triton, Larissa, and Proteus.

TRITON'S SURFACE
The dark areas (bottom left) are thought to be the result of geyser-like activity.

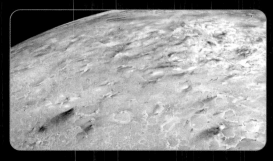

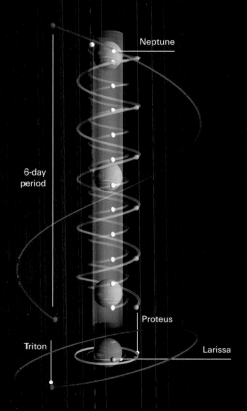

Neptune

6-day period

Proteus

Triton

Larissa

Pluto

The ninth and last planet is Pluto. It is very small and icy, unlike either the rocky planets or the gas giants. Because of its unique features, scientists think Pluto is a different kind of solar system object—an icy planetesimal, typical of the outer solar system.

CROSS SECTION

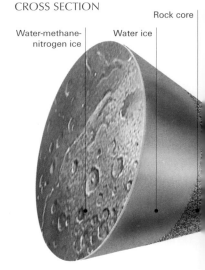

Rock core

Water-methane-nitrogen ice

Water ice

SIZE COMPARISON

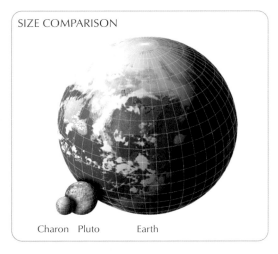

Charon   Pluto   Earth

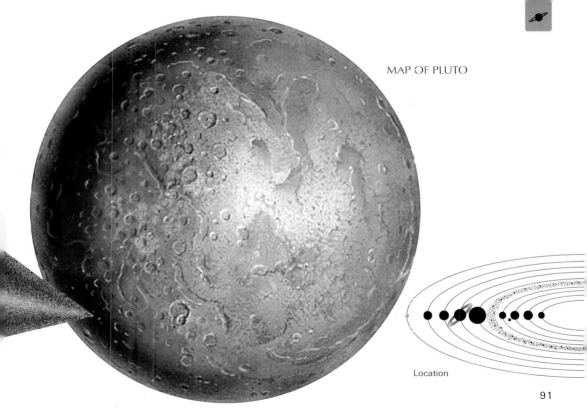

MAP OF PLUTO

Location

91

Pluto's Moon

Charon, Pluto's moon, is a mystery. Its surface seems entirely covered in water ice with none of Pluto's methane or nitrogen. It probably has a rocky core.

PLUTO AND CHARON
Charon is relatively large—its diameter is nearly half Pluto's. Scientists think that Pluto and Charon may be a double-planet system, which formed after a catastrophic collision.

CHARON

PLUTO'S DAY AND MONTH

Pluto and Charon keep the same faces turned toward each other. Charon revolves around Pluto in exactly the time that Pluto rotates on its axis. So Pluto's day and its month have the same length—about 6.4 Earth days.

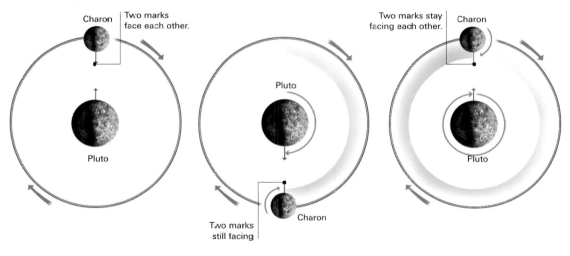

Start of Pluto's day and month

Halfway through rotation and orbit cycles

Rotation and orbit cycles completed

Pluto and Beyond

Pluto orbits the Sun in a region known as the Kuiper Belt. This area is home to icy asteroid-like bodies that are the leftovers from the formation of the solar system. Scientists think Pluto may have originally come from the Kuiper Belt.

FAR-DISTANT WORLDS
This Hubble Space Telescope image is our best view of Pluto (above) and Charon (below) so far.

PLUTO-KUIPER EXPRESS
No spacecraft has yet visited Pluto, but the Pluto-Kuiper Express is planned. After passing Pluto, the spacecraft will head into the Kuiper Belt.

THE OUTER REACHES
This cutaway view shows the planet's orbit and the Kuiper Belt. Pluto's elliptical orbit tilts 17 degrees to the orbits of the other planets.

Kuiper Belt | Pluto | Pluto's orbit

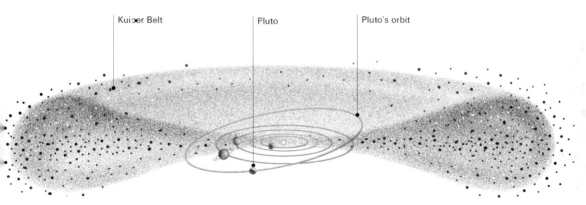

Comets

Comets are lumps of ice mixed with dust. Most exist beyond
Neptune, but sometimes gravity pulls a comet toward the inner
solar system. As the comet nears the Sun, its icy layers heat up
and boil away, forming a cloud of gas and dust called a coma.

A COMET ORBIT

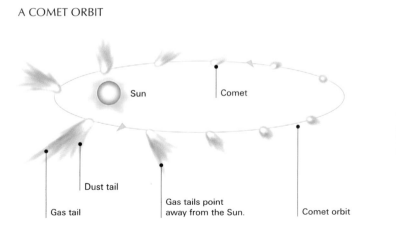

Dust tail

Gas tail

Gas tails point
away from the Sun.

Sun

Comet

Comet orbit

Nucleus

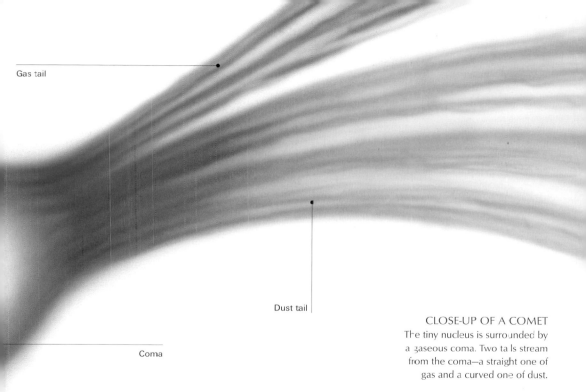

Gas tail

Dust tail

Coma

CLOSE-UP OF A COMET

The tiny nucleus is surrounded by a gaseous coma. Two tails stream from the coma—a straight one of gas and a curved one of dust.

Comet Watching

Short-period comets such as Halley and Encke orbit the Sun in 200 years or less and travel among the planets. Those with an orbit of more than 200 years are long-period comets. Comet Hale-Bopp will not return for another 2,400 years.

COMET HALE-BOPP
The most recent bright comet was Hale-Bopp, which swept the Northern Hemisphere skies in March and April of 1997.

COMET HALLEY
Probably the most famous comet is Halley. It appears every 75 or 76 years and last flew past Earth in 1986.

Planet Factfile

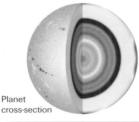

Planet
cross-section

SURFACE OF VENUS

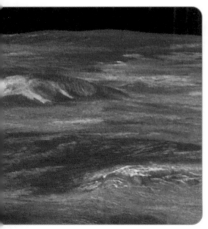

SUN

Rotation time: 25 Earth days at equator, 34 Earth days near poles
Diameter: 865,000 miles (1,392,000 km)
Mass: 333,000 x Earth
Volume: 1.3 million x Earth
Surface temperature: 9,900°F (5,500°C)
Core temperature: 27,900,000°F (15,500,000°C)
Surface gravity: 28 x Earth
Number of planets: 9

MERCURY

Distance from Sun: 36 million miles (58 million km)
Length of year: 88 Earth days
Rotation time: 59 Earth days
Solar day: 176 Earth days
Diameter: 3,029 miles (4,875 km)
Mass: 0.055 x Earth
Mean surface temperature: 800°F (430°C)
Surface gravity: 0.38 x Earth
Number of moons: none

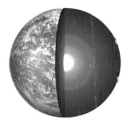

VENUS

Distance from Sun: 67 million miles (108 million km)

Length of year: 225 Earth days

Rotation time: 243 Earth days

Solar day: 117 Earth days

Diameter: 7,521 miles (12,104 km)

Mass: 0.82 x Earth

Mean surface temperature: 900°F (480°C)

Surface gravity: 0.90 x Earth

Number of moons: none

EARTH

Distance from Sun: 93 million miles (150 million km)

Length of year: 365.25 days

Rotation time: 23 hours 56 min

Solar day: 24 hours

Diameter: 7,926 miles (12,756 km)

Mass: 1.3×10^{24} lb (6×10^{24} kg)

Mean surface temperature: 72°F (22°C)

Surface gravity: 32.19 feet/second² (9.81 m/s²)

Number of moons: 1

MARS

Distance from Sun: 142 million miles (228 million km)

Length of year: 687 Earth days

Rotation time: 24 hours 37 min

Solar day: 24 hours 40 min

Diameter: 4,221 miles (6,794 km)

Mass: 0.11 x Earth

Mean surface temperature: -10°F (-23°C)

Surface gravity: 0.38 x Earth

Number of moons: 2

Planet
cross-section

JUPITER

Distance from Sun: 483 million miles (778 million km)

Length of year: 11.9 Earth years

Rotation time: 9 hours 55 min

Solar day: same as rotation time

Diameter: 89,405 miles (143,884 km)

Mass: 317.8 x Earth

Mean temperature at cloudtops: −240°F (−150°C)

Surface gravity: 2.3 x Earth

Number of moons: at least 61

SATURN

Distance from Sun: 890 million miles (1,432 million km)

Length of year: 29.4 Earth years

Rotation time: 10 hours 39 min

Solar day: same as rotation time

Diameter: 74,898 miles (120,536 km)

Mass: 95.2 x Earth

Mean temperature at cloudtops: −110°F (−80°C)

Surface gravity: 1.16 x Earth

Number of moons: at least 31

URANUS

Distance from Sun: 1,784 million miles (2,871 million km)

Length of year: 84.1 Earth years

Rotation time: 17 hours 14 min

Solar day: same as rotation time

Diameter: 31,763 miles (51,118 km)

Mass: 14.5 x Earth

Mean temperature at cloudtops: −355°F (−215°C)

Surface gravity: 1.17 x Earth

Number of moons: at least 27

NEPTUNE

Distance from Sun: 2,795 million miles (4,498 million km)
Length of year: 164.9 Earth years
Rotation time: 16 hours 7 min
Solar day: same as rotation time
Diameter 30,778 miles (49,532km)
Mass: 17.1 x Earth
Mean temperature at cloudtops: −265°F (−220°C)
Surface gravity: 1.77 x Earth
Number of moons: at least 13

PLUTO

Distance from Sun: 3,675 million miles (5,914 million km)
Length of year: 248 Earth years
Rotation time: 6.4 Earth days
Solar day: same as rotation time
Diameter: 1,430 miles (2,300 km)
Mass: 0.002 x Earth
Mean surface temperature: −380°F (−230°C)
Surface gravity: 0.06 x Earth
Number of moons: 1

JUPITER AND ITS MOON IO

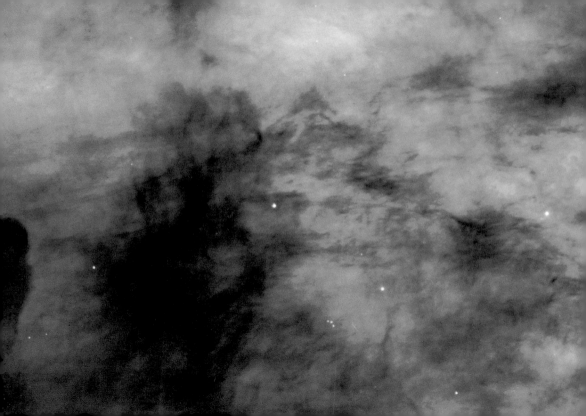

UNDERSTANDING
THE UNIVERSE

Stars

A star is a big ball of extremely hot gas—mostly hydrogen (about 90 percent) and helium. Nuclear fusion reactions in its core generate outward pressure that balances gravity and keeps the star shining for up to tens of billions of years.

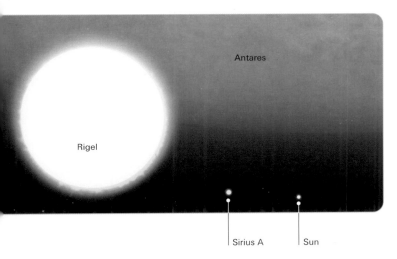

Antares

Rigel

Sirius A

Sun

SMALL STARS

The most common stars, red dwarfs such as Proxima Centauri, are smaller and cooler than the Sun. White dwarfs such as Sirius B are even smaller. The smallest are neutron stars—too tiny to be seen on the scale of this diagram.

LARGE STARS

Sirius A is a similar size to the Sun, but is hotter and therefore bluer. Rigel is both brighter—about 150,000 Suns—and larger. But the red supergiants, like Antares, dwarf even this.

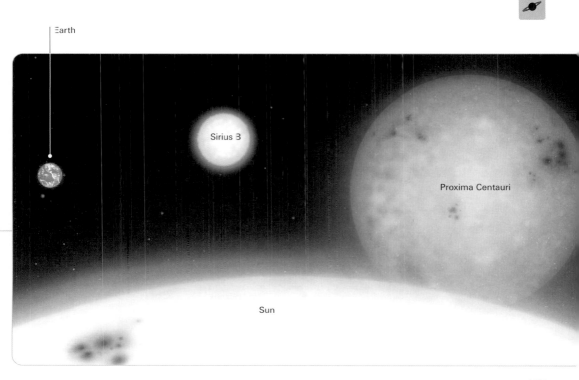

Earth

Sirius B

Proxima Centauri

Sun

Star Clusters

Most stars occur in groups or associations called clusters. Stellar associations contain relatively few stars across a huge space. Open clusters may have hundreds of stars in a smaller area. Globular clusters are more concentrated still, with up to a million stars.

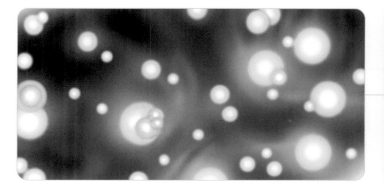

GLOBULAR CLUSTERS
A young globular cluster, at right, has many white-hot stars, plus dim red dwarfs and yellow stars like the Sun. As it ages, the stars evolve into red giants and then white dwarfs, far right.

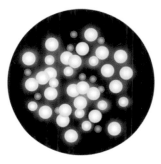
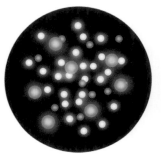

THE PLEIADES
This beautiful open cluster, left, is in the constellation of Taurus. It lies 375 light-years away and contains several hundred stars, including many young, hot blue stars.

OPEN CLUSTERS
Open clusters, such as this Quintuplet cluster, shine more brightly than other clusters. Over time, the stars may be scattered by the galaxy's gravitational forces. Eventually, the cluster itself can be hard to discern.

Other Solar Systems

The Sun isn't the only star with orbiting planets. Since 1995, more than 100 other planets have been discovered around stars similar to the Sun. Astronomers think they will eventually find that lots of stars have their own families of planets.

No shift = galaxy at rest

DOPPLER SHIFT
To find new planets and galaxies, astronomers look for tiny movements from a star. This is indicated by subtle changes in the color of the star's light, recorded by a method known as the Doppler shift.

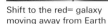 

Shift to the red= galaxy moving away from Earth

Shift to the blue = galaxy approaching Earth

 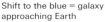

PLANET COMPARISON

So far, all the planets found around other stars are large bodies with masses similar to Jupiter's. Telescopes such as the COROT space telescope will help in the search for these planets.

The Life Cycle of a Star

A star is born from a lump of dusty gas in a nebula. The more massive a star is, the hotter it becomes and the faster its life runs. At the end of its life, millions or billions of years later, it will throw some of its material back into space. This sequence shows the life cycle of an average-sized star like the Sun.

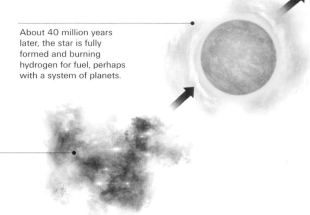

About 40 million years later, the star is fully formed and burning hydrogen for fuel, perhaps with a system of planets.

An ordinary star like the Sun is born in a nebula when a cloud of gas and dust starts to collapse and grow hotter.

After another 11 billion years, the star runs low on fuel. It swells and cools into a red giant that starts to pulsate.

About 1.5 billion years later the pulsations grow stronger. The bloated star sheds its outer layers and forms a planetary nebula.

In less than 50,000 years, the planetary nebula disperses into space. It leaves behind a shining, fading ember known as a white dwarf.

Variable Stars

While many stars shine steadily year after year, others seem to change in brightness. Giants and supergiants regularly grow larger and smaller, changing color and brightness—these are called pulsating variables.

Another kind of variable star is called an eclipsing binary. When one star passes behind the other, its light is blocked and the system seems to dim.

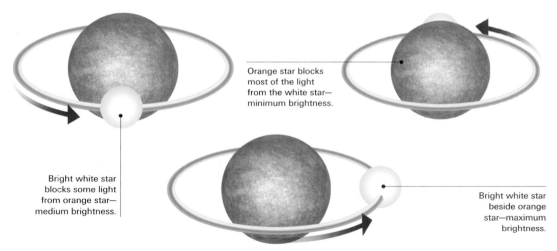

Orange star blocks most of the light from the white star— minimum brightness.

Bright white star blocks some light from orange star— medium brightness.

Bright white star beside orange star—maximum brightness.

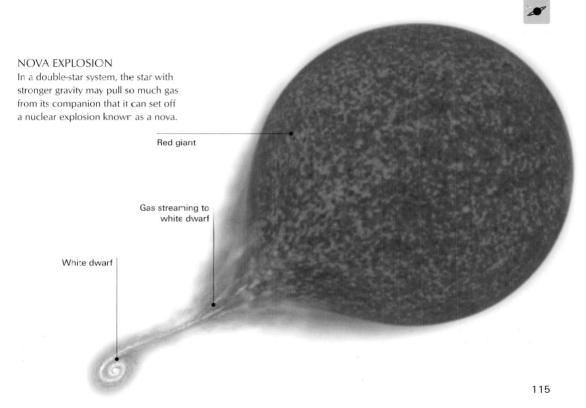

NOVA EXPLOSION
In a double-star system, the star with stronger gravity may pull so much gas from its companion that it can set off a nuclear explosion known as a nova.

Red giant

Gas streaming to white dwarf

White dwarf

Supernovas

A supernova is the catastrophic death of a star.
It can briefly outshine all other stars in its galaxy.
A white dwarf in a binary system and massive
stars can both end in this spectacular fashion.

SUPERNOVA REMNANT
About 120,000 years ago, a massive star
exploded in the constellation of Vela.
The explosion sent gas and dust into
space, where they will help to build new stars.

Supergiant star

Blue-white star

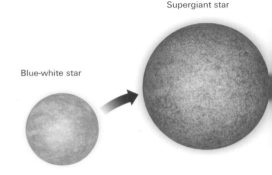

MASSIVE-STAR SUPERNOVA
A star at least eight times more massive than
the Sun will start life as a blue-white star. As
it runs out of fuel, its core collapses and the
star explodes in a supernova. A neutron star
or a black hole will be left.

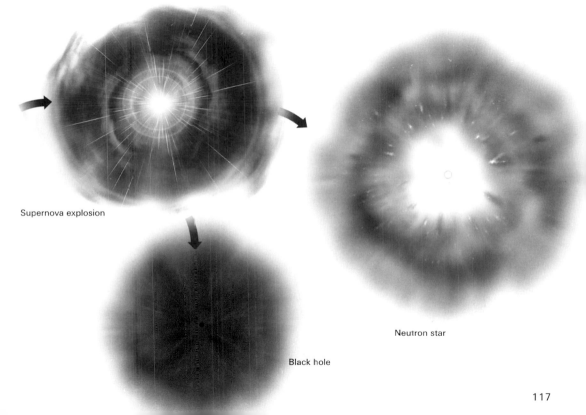

Supernova explosion

Neutron star

Black hole

117

Nebulas

Nebulas are thick clouds of dust and gas. In some, gravity packs the dust and gas so tightly that parts of it condense into stars. In others—planetary nebulas—we see the dying stages of a star.

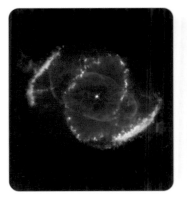

CAT'S-EYE PLANETARY NEBULA
When a star grows old, it throws off its outer layers. The layers become a shell of gas, lit by the now-exposed core.

GLOWING EMISSION NEBULA
This emission nebula surrounds bright newborn stars. The stars' ultraviolet radiation lights up the clouds.

DOUBLE BUBBLE NEBULA
This reflection nebula consists of hot, young stars inside a larger nebula. It is lit by the reflection of nearby stars.

CONE NEBULA
This dark cloud of dust and gas is lit from behind by a bright emission nebula. The energy from newborn stars causes these striking shapes.

Inside a Nebula

Nebulas are usually cold and do not shine.
An emission nebula, however, absorbs energy
from nearby stars, causing its gases to glow.

THE STAR-LIT EAGLE EMISSION NEBULA

PILLARS OF DUST AND GAS INSIDE THE NEBULA

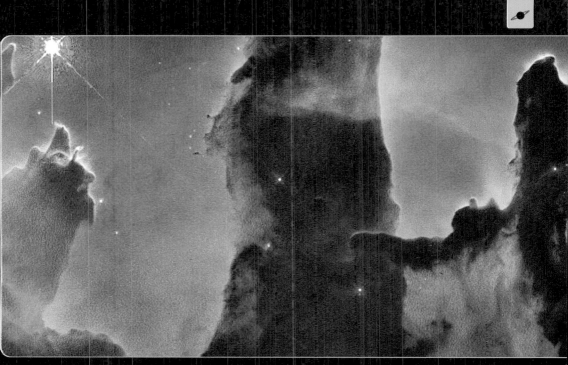

DENSER GLOBULES OF DUSTY GAS EMERGING FROM THE PILLARS

The Milky Way

From Earth, the Milky Way appears like a milky band of light. In fact, it is a spiral galaxy—our galaxy—seen from the inside. It is truly vast, more than 100,000 light-years in diameter.

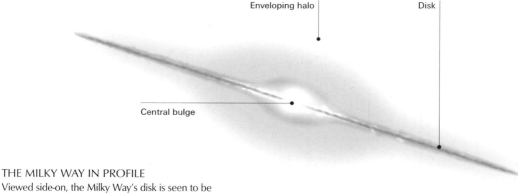

Enveloping halo

Disk

Central bulge

THE MILKY WAY IN PROFILE
Viewed side-on, the Milky Way's disk is seen to be much thinner than its central bulge. Surrounding it is a halo of globular clusters and old red stars.

OUTSIDE VIEW

Young, hot stars shine within open clusters along the Milky Way's spiral arms. The galactic bulge is a huge collection of older redder stars, and at its center lies a supermassive black hole.

Inside the Milky Way

The Milky Way contains stars, clusters, and gas clouds. All the prominent objects in our night sky come from a small part of our galaxy.

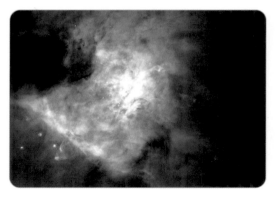

ORION NEBULA
The Orion nebula is part of a giant cloud of gas and dust—a "star factory" that has been at work for 12 million years. It glows due to the radiation emitted by hot stars within.

THE GALACTIC CENTER
This image, using infrared wavelengths, reveals stars normally hidden by thick dust in the Milky Way. It shows a central core, about 25,000 light-years away, made red by the dust.

OMEGA CENTAURI
This great globular cluster is in the constellation Centaurus.
It can be seen with the naked eye as a "woolly" star but
closer inspection reveals a dense, star-rich core.

Kinds of Galaxies

Galaxies are truly vast: incredible
islands of stars. Astronomers
classify them by shape into
three basic patterns—spirals,
ellipticals, and irregulars.

Spiral galaxy

SPIRAL GALAXY
These flat disk galaxies may contain
up to 10 trillion solar masses. The
arms mark out waves of star formation
sweeping around the galaxy.

Elliptical galaxy

Irregular galaxy

OTHER GALAXIES

Barred spirals are a subclass of the spirals. A "bar" emerges from the central bulge, and their arms emerge from the ends. Ellipticals include dwarf ellipticals and giant ellipticals, which are mergers of galaxies. Galaxies lacking in obvious structure are called irregular.

Barred spiral galaxy

Galaxy Shapes

Spiral and barred spiral galaxies might be the most immediately recognizable of galaxy shapes, but a range of other shapes exist, from circles and ellipsoids to shapeless blobs.

A NEARLY PERFECT RING
This galaxy, known as Hoag's Object, has hot, blue stars circling around the yellow nucleus of older stars.

POLAR RING GALAXY
This rare type of galaxy features an old central group of stars and a young ring of stars rotating farther out.

TIGHTLY WOUND
This barred spiral galaxy, NGC 2787, has arms of dust tightly wrapped around the bright nucleus.

THE WHIRLPOOL GALAXY, M51
This classic spiral, seen face-on, was one of the first spiral
galaxies to be identified. This photo from the Hubble Space
Telescope shows a close-up of the central regions.

Galactic Collisions

Astronomers are discovering that the Universe is a dynamic place. Galaxies may evolve through colliding and merging with other galaxies. In particular, giant elliptical galaxies are thought to be the result of collisions.

STAR SYSTEMS MERGE
The collision of two galaxies has left Messier 64 with a dark band of dust around its nucleus. This appearance has earned it the nickname of Black Eye.

DID YOU KNOW?
Collisions explain why giant ellipticals are so big: they swallow their neighbors.

SPINNING SIDEWAYS
This image shows a spiral galaxy sliding through the larger galaxy, NGC 1275. The bright blue spots are areas of active star formation.

TWO GALAXIES COLLIDE
When galaxies collide, gravitational pull distorts their shapes. Gas and dust are often pulled from both galaxies and stretched into long streams of celestial matter in which stars form.

Active Galaxies

An object so dense that not even light can escape its gravity is called a black hole. Black holes form after a supernova explosion. Some galaxies—known as active galaxies—have a massive black hole at their core that feeds on entire stars.

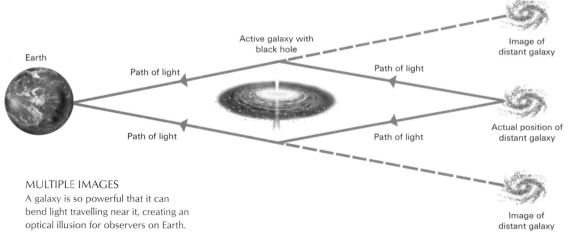

Earth

Active galaxy with black hole

Path of light

Path of light

Path of light

Path of light

Image of distant galaxy

Actual position of distant galaxy

Image of distant galaxy

MULTIPLE IMAGES
A galaxy is so powerful that it can bend light travelling near it, creating an optical illusion for observers on Earth.

GALACTIC BLACK HOLE

Active galaxies produce a huge amount of energy in a small space. Jets of material emerge from a disk of gas surrounding a supermassive black hole. As the gas falls into the hole, it becomes very hot and emits energy.

The Local Group

Galaxies occur in clusters—even clusters of clusters of incredible size—bound together by gravity. Our galaxy is part of the Local Group, a small grouping of at least 35 galaxies. Rich clusters contain thousands of galaxies.

NGC 6822, irregular

IC 1613, irregular

Andromeda (M31), spiral

M32, elliptical

Pinwheel (M33), spiral

NGC 147, dwarf elliptical

NGC 185, dwarf elliptical

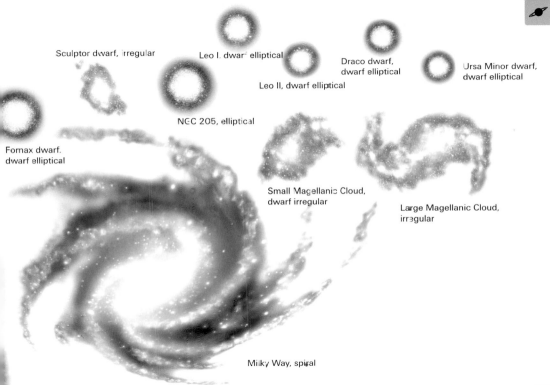

Sculptor dwarf, irregular

Leo I, dwarf elliptical

Leo II, dwarf elliptical

Draco dwarf, dwarf elliptical

Ursa Minor dwarf, dwarf elliptical

NGC 205, elliptical

Fornax dwarf, dwarf elliptical

Small Magellanic Cloud, dwarf irregular

Large Magellanic Cloud, irregular

Milky Way, spiral

Galaxies of the Local Group

The Local Group has a lot of small member-galaxies and a few big, bright ones such as the Andromeda and the Milky Way. The Pinwheel is the third-largest galaxy in the group.

ANDROMEDA GALAXY
This galaxy is bright enough to be visible to the naked eye—even though it is about 2.5 million light-years away. This image shows the galaxy's halo.

SUPERNOVA 1987A
On February 24, 1987, an exploding star in the Large Magellanic Cloud gave modern astronomers their first close-up view of a supernova.

The Birth of Galaxies

Astronomers believe that most galaxies must have formed before 10 billion years ago—within a few billion years of the Big Bang. But they are less sure of how they formed. Two scenarios have been suggested, and it's possible that both were at work.

BOTTOM-UP SCENARIO
Galaxies may have assembled themselves in stages, over a period of billions of years. "Building blocks" of material may have come from surrounding globular clusters.

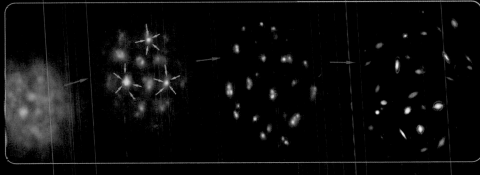

TOP-DOWN SCENARIO
A second theory suggests that the early Universe consisted of large clouds of gas. These vast structures broke up over time, creating galaxy clusters and superclusters, then galaxy-sized chunks and, later, stars.

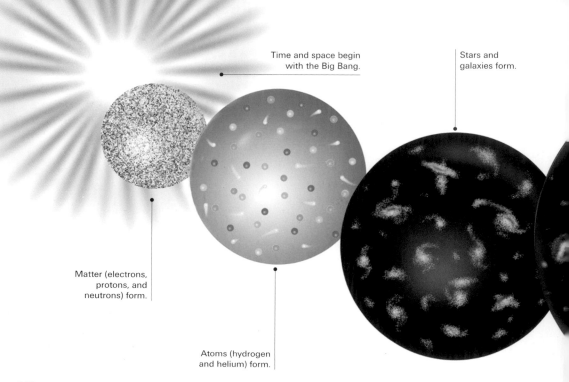

Time and space begin
with the Big Bang.

Stars and
galaxies form.

Matter (electrons,
protons, and
neutrons) form.

Atoms (hydrogen
and helium) form.

140

The Big Bang

The Universe was born almost 14 billion years ago in a gigantic explosion that astronomers call the Big Bang. For a fraction of a second, it was a super-hot, super-compressed speck of exotic particles. As it expanded and cooled, familiar particles formed. Then, atoms formed, later clumping together under gravity's influence to form stars and galaxies.

Today's Universe
has perhaps
50 billion galaxies.

DID YOU KNOW?
Astronomers who study the structure of the Universe are called cosmologists.

The Future of the Universe

The Universe is still expanding, but it is not known if it will always do so. The Universe is expected to expand indefinitely, but it may end in a Big Crunch.

INDEFINITE EXPANSION

If the Universe keeps expanding, galaxies will drift apart and eventually die out. Ordinary matter will disintegrate, leaving a boundless "sea" of particles.

Galaxies die out.

Matter disintegrates.

Universe at present

Galaxies drift farther apart.

THE BIG CRUNCH

If there is enough matter in the Universe, at some point the force of gravity will bring expansion to a halt and then reverse, ending as a Big Crunch. Recent research suggests this scenario is unlikely.

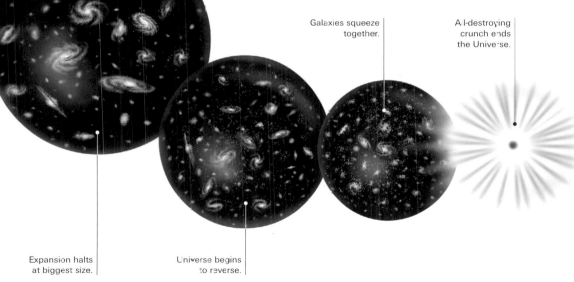

Galaxies squeeze together.

All-destroying crunch ends the Universe.

Expansion halts at biggest size.

Universe begins to reverse.

Records of the Universe

Hottest planet surface in the solar system
The surface of Venus, 900°F (480°C).

Coldest recorded surface in the solar system
Triton, the largest satellite of Neptune, with a temperature of -391°F (-235°C).

Biggest crater in the solar system
The Aitken Basin on the Moon's south pole, 1,600 miles (2,500 km) in diameter.

Tallest mountain in the solar system
Olympus Mons on Mars, rising 15 miles (24 km) above its base.

Biggest canyon in the solar system
Valles Marineris on Mars, about 2,500 miles (4,000 km) long, with a maximum width of 370 miles (600 km) and a maximum depth of 5 miles (8 km).

Largest planet in the solar system
Jupiter, with 317.8 times the mass of Earth, and about 11 times its diameter.

Largest satellite in the solar system
Jupiter's Ganymede, 3,273 miles (5,268 km) in diameter. It is larger than either Mercury or Pluto.

Greatest meteor shower
The Leonids on November 13, 1833, with up to 200,000 meteors per hour.

Largest meteorite
Hoba meteorite in Namibia, weighing 65 tons (60 tonnes)—about as heavy as nine elephants!

Largest asteroid
1 Ceres, 567 miles (913 km) in diameter.

Closest comet to Earth
Comet Lexell in 1770 passed within 1.4 million miles (2.2 million km) of Earth—less than six times the distance to the Moon.

Longest comet tail
Great Comet of March 1843, 190 million miles (300 million km) long. This tail was long enough to reach from the Sun to well past the orbit of Mars.

Most massive star
Eta Carinae, approximately 150 times as massive as the Sun. Astronomers are not certain if Eta Carinae is one star or two.

Least massive star
Gliese 105C, about 10 percent as massive as the Sun.

Star with largest diameter in the night sky
Betelgeuse in Orion, about 800 times the Sun's diameter.

Nearest star
Proxima Centauri, third member of the Alpha Centauri system. This cool red dwarf star lies about 4.2 light-years away, about 0.1 light-year closer to us than the other two stars in the system.

Globular star cluster with the most stars
Omega Centauri, with 1.1 million stars. This globular star cluster measures about 180 light-years in diameter.

Most massive nearby galaxy
Giant elliptical M87 in the constellation of Virgo, with at least 800 billion Suns' worth of mass.

Least massive galaxy known
The Pegasus II dwarf elliptical, about 10 million solar masses

Nearest galaxy
The Canis Major dwarf galaxy in Canis Major is 25,000 light-years away from the solar system and 42,000 light-years from the Milky Way's center.

OLYMPUS MONS ON MARS

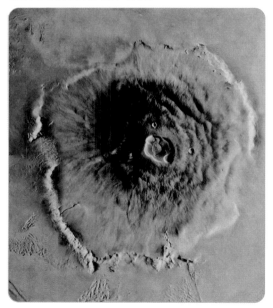

145

SPACE EXPLORATION

Looking into Deep Space

Astronomical research presents a fascinating window onto outer space. Today, our globe is dotted with hundreds of observatories that feature telescopes with mirrors up to 33 feet (10 m) across and radio dishes that are even larger. In space, too, telescopes continue to reveal ever-greater marvels from outer space.

STAR FACTORY
This nebula is just visible to the naked eye. When seen in detail, a giant cloud of dust and hydrogen gas is revealed.

STUNNING UNIVERSE
With each new development in technology, astronomers unlock more secrets of the Universe and how it works.

Early Space Exploration

People have long dreamed of flying into space. It became a reality in the 1950s, with the Soviet Union and United States competing for glory.

SPUTNIK 1

In 1957 the Soviet Union launched the bleeping Sputnik 1, the world's first artificial satellite of Earth. Thus began the "space race" between the USA and the USSR.

GAGARIN'S TRIUMPH

Yuri Gagarin made history when he blasted off on April 12, 1961 in the Vostok 1 spacecraft. He was the first person to fly in space, completing an orbit of the Earth in 108 minutes.

APOLLO MISSIONS

Apollo 8 was the first craft to carry humans beyond Earth's
gravity. More Apollo test flights followed, culminating in
Apollo 11's historic landing of the first men on the Moon.

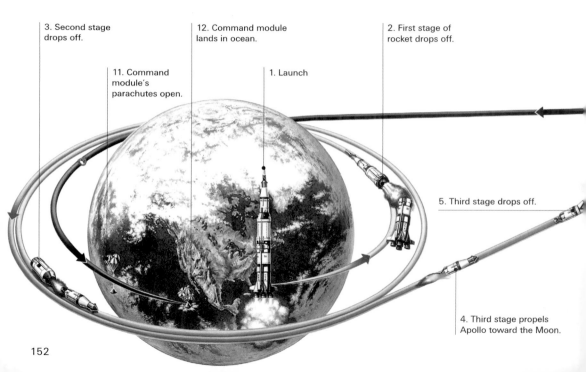

3. Second stage drops off.

12. Command module lands in ocean.

2. First stage of rocket drops off.

11. Command module's parachutes open.

1. Launch

5. Third stage drops off.

4. Third stage propels Apollo toward the Moon.

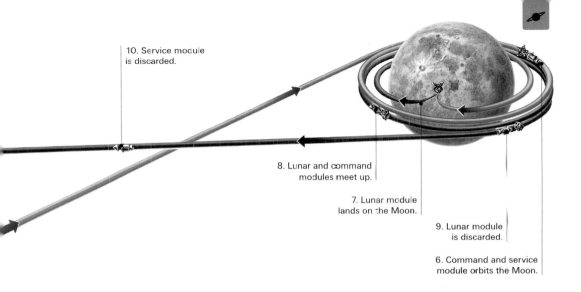

10. Service module is discarded.

8. Lunar and command modules meet up.

7. Lunar module lands on the Moon.

9. Lunar module is discarded.

6. Command and service module orbits the Moon.

Journey to the Moon

The Apollo spacecraft was launched into space by the powerful Saturn 5 rocket. Two astronauts landed on the Moon while a third orbited in the command and service modules.

Man on the Moon

In July 1969, Neil Armstrong became the first man on the Moon. Five other Apollo missions landed on the Moon, collecting rock samples.

FOOTPRINTS TO LAST
The 12 Apollo moonwalkers have left footprints on the Moon's surface that will last millions of years. Wind is non-existent and the only erosion comes from micrometeorites.

THE WORLD WATCHES
Millions saw these remarkable events on their televisions—American astronaut Edwin Aldrin of Apollo 11 stands beside his country's flag on the surface of the Moon.

Technological Developments

In the fifty years since the first artificial satellites were launched into space, spacecraft of all sizes, shapes, and purposes have followed. The result is a growing body of information early astronomers could only dream of.

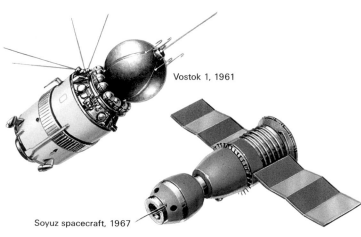

Vostok 1, 1961

Soyuz spacecraft, 1967

GEMINI SPACECRAFT
Among the first manned space modules, the American Gemini 3 successfully completed its first manned flight in 1965.

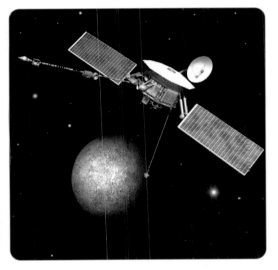

STUDYING THE PLANETS
Most of what we know about Mercury comes from just one spacecraft, Mariner 10, which made three flybys in 1974-75. Today, probes are able to study even the far-flung planets.

INTERNATIONAL EFFORT
Begun in 1998, the International Space Station marks a new age in cooperation between nations. When complete, it will hold as many as seven crew and their equipment.

The Space Shuttle

To get manned and unmanned craft into space you need a rocket. Today, the reusable American space shuttle and the three-stage Russian Soyuz rocket are the first choices for manned spacecraft.

Space shuttle in orbit

The external tank falls into the Pacific Ocean.

The solid rocket boosters shoot away.

Liftoff

LAUNCHING THE SPACE SHUTTLE
The shuttle is piggybacked to an external tank and two solid rocket boosters. Once the fuel is spent, the boosters drop away and are recovered; the tank is discarded.

A REUSABLE SHUTTLE
The space shuttle was built to carry telescopes and other satellites into orbit around Earth.

Upon reentry, the shuttle surface glows red hot.

Touchdown

Space Stations

Since the early 1970s, more than 100 men and women have worked above Earth on board various space stations. The stations come with life-support systems for their crews.

VISIT TO MIR
A space shuttle (foreground) is just about to dock with Mir. This Russian space station was visited a number of times before it left orbit in 2001.

SKYLAB SPACE STATION
Three separate crews manned Skylab during its flight, which lasted from May 1973 to July 1979. The station conducted a series of UV experiments and some X-ray studies of the Sun.

INTERNATIONAL SPACE STATION
When finished, this station will include more than 100 modules from 16 countries. It will be more than 100 yards (100 m) long and weigh more than 500 tons (450 tonnes).

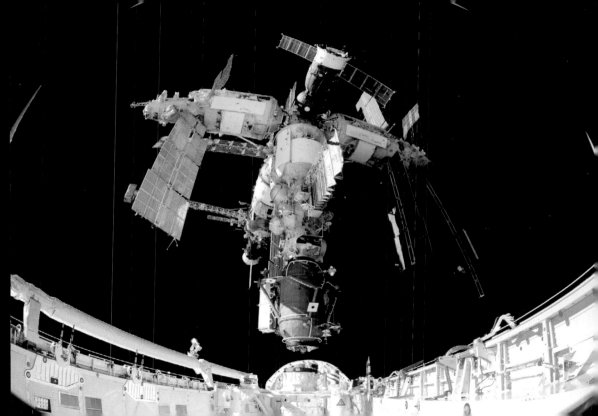

Working in Space

Astronauts may spend months on a space station, doing highly specialized work. They conduct experiments, make repairs, and study the effects on their bodies of being "weightless."

> DID YOU KNOW?
> **Astronauts grow 1–2 inches (2.5–5 cm) taller while in space. When they return to Earth, the astronauts shrink back to their preflight size.**

EVERYDAY TASKS
An astronaut stores a cordless power tool needed to keep the Hubble Space Telescope in perfect working order.

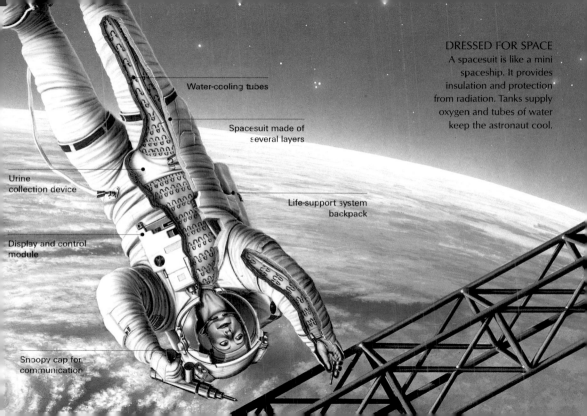

Water-cooling tubes

Spacesuit made of
several layers

Urine
collection device

Life-support system
backpack

Display and control
module

Snoopy cap for
communication

DRESSED FOR SPACE
A spacesuit is like a mini
spaceship. It provides
insulation and protection
from radiation. Tanks supply
oxygen and tubes of water
keep the astronaut cool.

Living in Space

While it sounds glamorous, life in space can be hard work for astronauts. They have to cope with "weightlessness," which affects the flow of blood, wastes away muscles, and weakens bones.

DID YOU KNOW?

Without gravity providing an anchor, astronauts can lose sense of up and down.

FLYING FREE
Backpacks with small thrusters can turn an astronaut into a mini-spaceship.

WEIGHTLESS PEDALING
Astronauts must exercise for hours every day to keep their bodies strong.

FLOATING INSIDE
Conditions inside a space craft are often cramped, and everything needs to be secured—including sleeping bags, food, and documents.

Probing Space

Since 1957, when the first artificial satellite was launched, probes have provided much valuable information on our solar system.

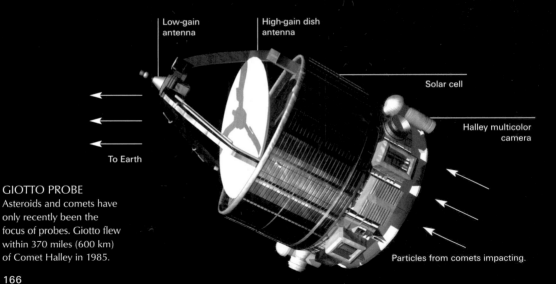

Low-gain antenna

High-gain dish antenna

Solar cell

Halley multicolor camera

To Earth

Particles from comets impacting.

GIOTTO PROBE
Asteroids and comets have only recently been the focus of probes. Giotto flew within 370 miles (600 km) of Comet Halley in 1985.

CLEMENTINE PROBE
The Moon is a favorite subject of probes. This one found possible evidence of water on the Moon.

Driving to Work

As the nearest planets to Earth, the terrestrials (Mercury, Venus, and Mars) were the first to attract the attention of probes. Mars in particular has been visited by Viking and, more recently, the Pathfinder, Spirit, and Opportunity rovers.

REMOTE STEERING
Mission scientists on Earth direct the movements of the first rover on Mars.

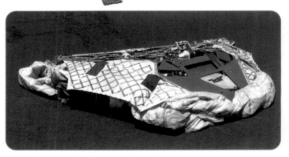

OPPORTUNITY'S EMPTY NEST
The Mars rover, Opportunity, has left its lander in quest of new discoveries.

Parachute opens, slowing descent of lander.

AIRBAGS LOWER SPACECRAFT

The 1997 Mars Pathfinder mission tested airbags as a way of lowering a spacecraft. Airbags do not contaminate the planet's surface. Opportunity and Spirit rovers also used this method.

Airbags soften the landing.

Airbags around lander inflate.

Airbags deflate.

Lander petals open.

Sojourner rover leaves lander and explores nearby.

Outer Planet Probes

Sending probes to the outer gas planets is no easy matter—the closest, Jupiter, is five times farther from the Sun than Earth.

CASSINI-HUYGENS PROBE
This probe to Saturn has two parts: the orbiter Cassini to study its rings, and the Huygens probe to study Saturn's largest moon, Titan.

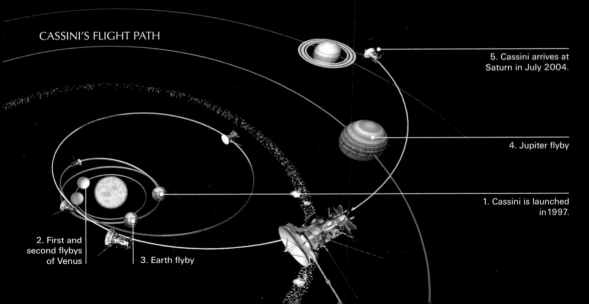

CASSINI'S FLIGHT PATH

5. Cassini arrives at Saturn in July 2004.

4. Jupiter flyby

1. Cassini is launched in 1997.

2. First and second flybys of Venus

3. Earth flyby

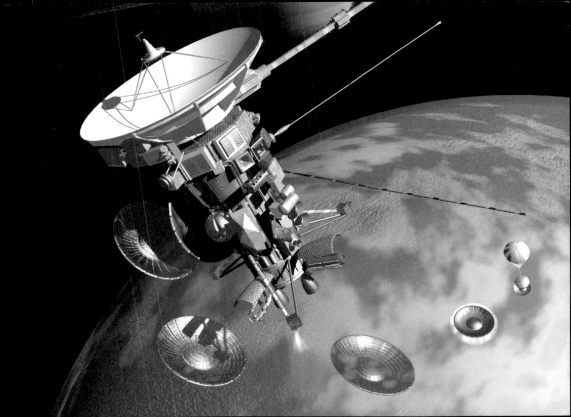

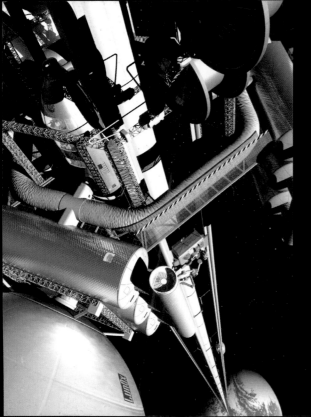

SPACE ELEVATOR

It may sound fantastic, but scientists are investigating a space elevator as a way of moving cargo from Earth to a stationary craft in space.

FUTURE CHALLENGES

This sleek X-33 was to be the first of a new generation of reusable launch vehicles. After a run of problems, however, it was cancelled in 2001.

The Future of Space Travel

New technologies hold out the promise of ever-more efficient and exciting space travel. Ion drive engines that run on electrically charged atoms, nuclear power, spacecraft with sails to catch sunlight, and space elevators are all being investigated.

DID YOU KNOW?

A space elevator might be a super-strong cable extending from Earth to a satellite almost 25,000 miles (40,000 km) above.

Searching for Life

Scientists think Mars, Jupiter's moon Europa, and Saturn's moon Titan are the best places to look for signs of life. They also scan radio signals from space for signs of alien life.

A MARTIAN LAKE?
A huge lake might once have filled the blue area in this image of Mars. Such areas are targeted in the search for life.

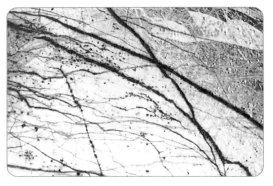

THE ICY CRUST OF JUPITER'S MOON
The red lines in this picture are cracks in Europa's icy crust. Water has filled the gaps and then frozen. Underneath may be a layer of water or soft ice where life might exist.

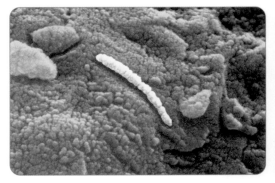

EVIDENCE OF EARLY MARTIAN LIFE?
A meteorite from Mars drew worldwide attention when scientists announced it contained evidence of Martian life. This included micro-structures (in yellow), similar to bacteria.

Invasion of Mars

Though it seems the stuff of science fiction, astronomers have considered whether there is intelligent life on other planets, and whether humans could live in outer space. Our nearest neighbor, Mars, has captured the imagination for both these reasons.

DID YOU KNOW?

Many people have reported UFO encounters, but reliable proof for UFOs is still lacking.

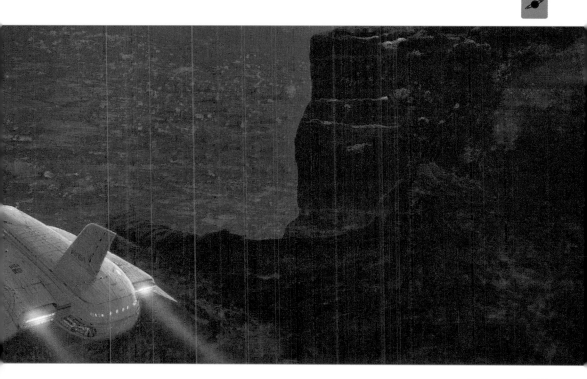

Space Exploration Timeline

Surveyor, 1966–1968

1957 Launch of Sputnik 1, the first artificial satellite, by the Soviet Union starts the Space Race.

1959 Photographs of the Moon's farside by Luna 3.

1961 The Soviet Union's Yuri Gagarin becomes the first man in space.

1962 Mariner 2 discovers Venus' heavy atmosphere and hot surface.

1965 Mariner 4 is the first spacecraft to fly past Mars.

1969 Neil Armstrong and Edwin Aldrin make the first manned landing on the Moon (Apollo 11).

1973 First flyby of Jupiter, by Pioneer 10.

1974 First close-up photos of Venus' cloud tops and Mercury's heavily cratered surface, by Mariner 10.

1975 First photos from the surface of Venus, by Venera 9.

1976 Viking 1 and 2 successfully land on Mars, in an attempt to detect life on the planet.

1979 Voyager 1 and 2 fly past Jupiter, discovering its rings. First flyby of Saturn, by Pioneer 11.

1980 First detailed study of Saturn, by Voyager 1.

1983 Infrared Astronomical Satellite (IRAS) surveys the infrared sky.

1986 First flyby of Uranus, by Voyager 2. Close flyby of Comet Halley, by Europe's Giotto spacecraft.

1989 First flyby of Neptune, by Voyager 2.

1990 Hubble Space Telescope launched. Magellan spacecraft maps Venus by radar.

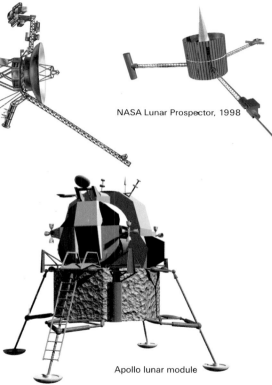

Luna 16, 1970

Voyager 2, 1981

NASA Lunar Prospector, 1998

1991 Galileo spacecraft on the way to Jupiter makes the first asteroid flyby, of 951 Gaspra. Launch of Compton Gamma-ray Observatory.

1995 Galileo spacecraft arrives at Jupiter.

1997 Mars Pathfinder spacecraft lands on Mars with the Sojourner rover.

1999 Launch of Chandra X-ray Satellite Observatory.

2003 The first manned Chinese spacecraft, Shenzhou 5, takes off from the Gobi Desert Launch of the Spitzer Space Infrared Telescope, NASA's fourth and final "Great Observatory."

2004 The orbiter Cassini arrives at Saturn. The Opportunity rover lands on Mars in January, 2004, three weeks after the Spirit rover lands.

Apollo lunar module

ASTRONOMY THROUGH THE AGES

Signs in the Sky

Early skywatchers studied the stars, kept records of the movements of the planets, and compiled calendars. The night sky seemed a magical place. They created shapes from the patterns of the stars (constellations), which they named after mythological characters. In the West, early constellations include Leo, Cancer, Gemini, and Taurus.

The First Skywatchers

The monuments of ancient civilizations often show that their builders had a great grasp of astronomy. Remarkably, they developed this knowledge independently of each other.

MAYAN PYRAMIDS
The windows of this pyramid at Palenque are arranged to reflect the rising and setting cycles of Venus.

STONEHENGE
Stonehenge may have been an early observatory. It was built in three main stages and completed around 1500 BC.

ARCHITECTURE GUIDED BY THE STARS

Around 3000 BC the Egyptians carved out the massive Giza pyramids. Some claim they are aligned with the belt stars in the constellation of Orion—known to the Egyptians as Osiris, God of the Dead.

ANCIENT BABYLONIAN CLAY TABLETS

Astronomers in Babylonia (today's Iraq) recorded the movements of stars and planets over 2,500 years ago.

The Zodiac

As Earth orbits the Sun, the Sun's path, projected onto our skies, traces a line known as the ecliptic. This line cuts through various constellations; 12 of them make up the zodiac.

NAMING THE ZODIAC
In the West, we know the 12 constellations of the zodiac by their ancient Greek names. Other cultures have their own zodiacal creatures. This image shows Asian religious characters surrounded by the zodiac.

CHINESE ASTRONOMY

This disk from the Tang Dynasty (AD 618–907) shows the solstices and equinoxes, their zodiacal cycle, and the phases of the Moon.

EGYPTIAN CONSTELLATIONS

Constellation figures such as this one decorate the tombs of Egyptian pharaohs.

The Beginnings of Astronomy

Ancient astronomers from Mesopotamia, Egypt, China, and elsewhere developed a sophisticated understanding of the movements of celestial objects. This knowledge was linked to religion and ritual. It was the ancient Greeks, with knowledge inherited from Mesopotamia, who began the first scientific study of the Universe.

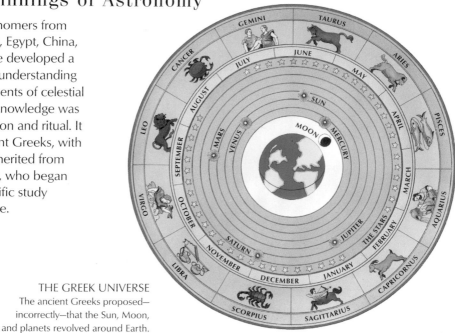

THE GREEK UNIVERSE
The ancient Greeks proposed—incorrectly—that the Sun, Moon, and planets revolved around Earth.

THE ROLE OF GODS

This engraving from an Egyptian mummy case shows Shu, the god of the atmosphere, lifting up his daughter Nut, goddess of the sky, to separate her from Earth.

Models of the Universe

Scientific advances in the sixteenth century saw many long-held ideas overturned. Of great interest was whether or not Earth was the center of the Universe.

NICOLAUS COPERNICUS
Astronomy was jolted awake in 1543, when Copernicus—on his deathbed—published his Sun-centered model of the Universe.

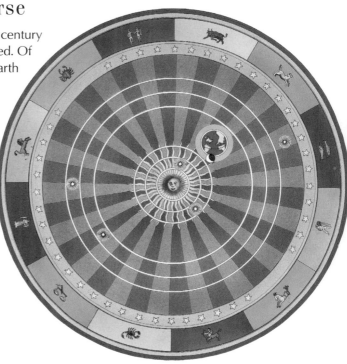

CLAUDIUS PTOLEMY

The model, described by Ptolemy in AD 150, asserted that all heavenly bodies orbited Earth. The idea was supported by the Church.

TYCHO BRAHE

Brahe (1546–1601) drew on the ideas of Copernicus and Ptolemy. In his model, the other planets orbit the Sun, but the Sun orbits Earth.

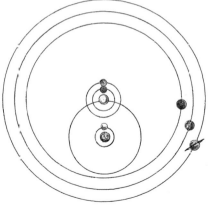

Scientific Breakthrough

Despite Church opposition, the theory of a Sun-centered Universe took hold. Support came from a new invention—the telescope—as well as from that most ancient of tools: observation.

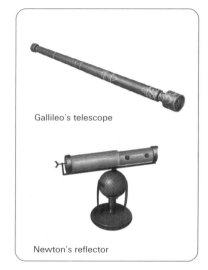

Gallileo's telescope

Newton's reflector

EARLY TELESCOPES
Galileo used telescopes that magnified 20 to 30 times. Newton designed a new type with mirrors. It gave a clearer view.

Gravity in action

GALILEO GALILEI (1564–1642)
Italian-born Galileo was the first person to report his use of the telescope for astronomical observation. He discovered previously invisible objects, including moons around Jupiter.

ISAAC NEWTON (1642–1727)
Newton was a genius whose scientific theories spanned many fields. He proposed a theory of gravity, explaining the mechanics of both the solar system and the everyday world.

A Great Leap Forward

Since the 1700s, new theories and technologies have led to big increases in our understanding of the Universe. The most important development was the spectograph, invented in the 1850s.

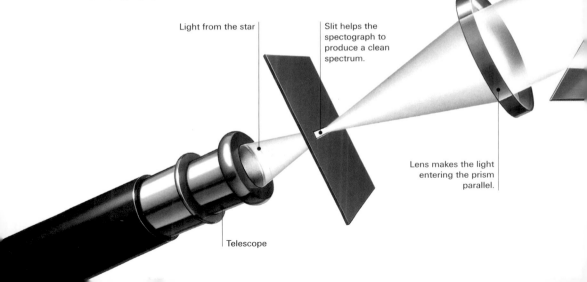

Light from the star

Slit helps the spectograph to produce a clean spectrum.

Lens makes the light entering the prism parallel.

Telescope

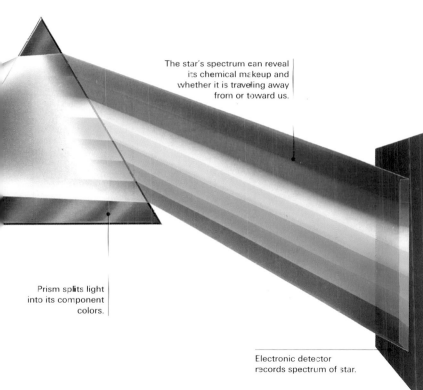

The star's spectrum can reveal its chemical makeup and whether it is traveling away from or toward us.

Prism splits light into its component colors.

Electronic detector records spectrum of star.

Optical Astronomy

The first telescopes were refractors, but astronomers soon learned how to make much bigger optical telescopes using mirrors. The Keck telescopes on Mauna Kea, Hawaii, are the world's largest.

Keck II dome

The telescope is mounted on a moveable base to track the movements of stars and galaxies.

Operators in the control room move and point the telescope.

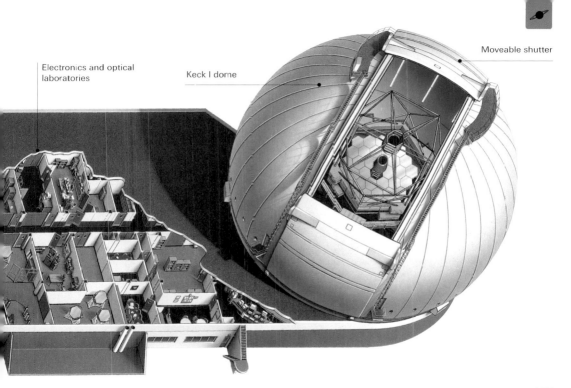

Electronics and optical laboratories

Keck I dome

Moveable shutter

Radio Astronomy

Traditional telescopes are optical, which means they collect the kind of light our eyes can see. Planets, stars, and galaxies also send out invisible radiation, such as radio waves. This type of radiation penetrates our atmosphere relatively easily, so can be studied from Earth.

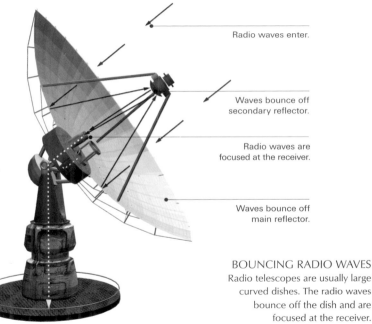

Radio waves enter.

Waves bounce off secondary reflector.

Radio waves are focused at the receiver.

Waves bounce off main reflector.

BOUNCING RADIO WAVES
Radio telescopes are usually large curved dishes. The radio waves bounce off the dish and are focused at the receiver.

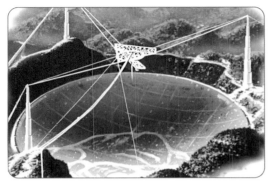

ARECIBO RADIO TELESCOPE

This enormous radio telescope is set in a hollow on the island of Puerto Rico. It studies planets and stars, and also listens for radio signals that might come from extraterrestrial life.

BIG EARS ON THE COSMOS

Radio telescopes are much larger than optical ones. The Very Large Array in New Mexico, USA, combines 27 dishes to form a single large telescope.

DID YOU KNOW?

The Arecibo antenna measures 1,000 feet (305 m) across. It is the largest single radio telescope.

Astronomy from Space

Earth's atmosphere both distorts and blocks parts of the electromagnetic spectrum. The solution: put telescopes into outer space.

INFRARED SPACE
OBSERVATORY

This satellite investigated water clouds in distant galaxies.

IRAS SPACE TELESCOPE

The InfraRed Astronomy Satellite surveyed the dust of the Milky Way.

COSMIC BACKGROUND
EXPLORER (COBE)

This satellite studied the radiation left over from the Big Bang.

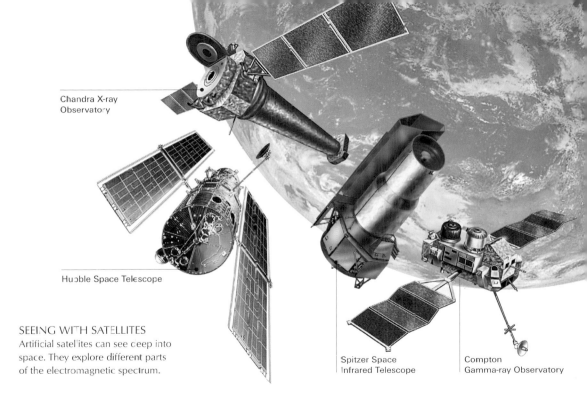

Chandra X-ray
Observatory

Hubble Space Telescope

Spitzer Space
Infrared Telescope

Compton
Gamma-ray Observatory

SEEING WITH SATELLITES

Artificial satellites can see deep into
space. They explore different parts
of the electromagnetic spectrum.

The Hubble Space Telescope

With a main mirror measuring only 8 feet (2.4 m) across, the Hubble Telescope is not large. But the images received from it have made it the most famous telescope of all.

A SEA OF BLUE
Launched in April 1990, the Hubble Space Telescope floats 380 miles (600 km) skyward. Floating above the blue planet, it views the sky 24 hours a day.

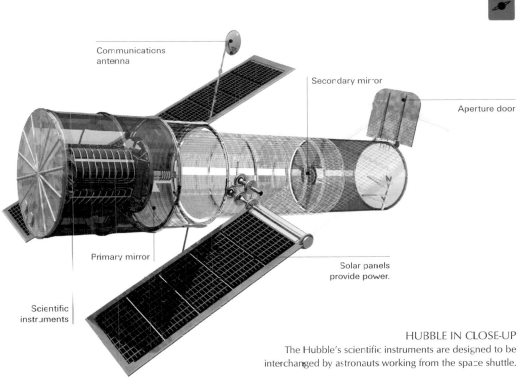

Communications antenna

Secondary mirror

Aperture door

Primary mirror

Solar panels provide power.

Scientific instruments

HUBBLE IN CLOSE-UP
The Hubble's scientific instruments are designed to be interchanged by astronauts working from the space shuttle.

Timeline of Astronomy

Leo

Ptolemy

Galileo's telescope

30,000 BC Lunar phases scratched on bone—the oldest astronomical record?

4000 BC Sumerians of Mesopotamia make the first records of Leo, Taurus, and Scorpius, the oldest constellations still used today.

600 BC Greek philosopher Thales probably knows the cause of solar and lunar eclipses.

350 BC Greek philosopher Aristotle provides a scientific explanation of why Earth is round.

325 BC Greek mathematician Eudoxus explains celestial motions, putting Earth at the center of the Universe. 25 years later, astronomer Aristarchus proposes a Sun-centered model. This theory is ignored until Copernicus' time.

AD 150 Greek astronomer Claudius Ptolemy publishes a detailed summing-up of all the ancient world's astronomical knowledge. It dominates for more than 1,000 years.

AD 165 Chinese astronomers make the first accurately dated observations, recording sunspots on the Sun's face.

1543 Nicolaus Copernicus publishes his book *On the Revolutions of the Celestial Spheres*, proposing a Sun-centered solar system.

1576 Tycho Brahe begins compiling the world's most accurate observations of the motions of stars.

1609 Galileo Galilei builds a telescope, based on the design of Hans Lippershey's telescope, built in 1608.

1619 Johannes Kepler publishes the last of his three laws of planetary motion.

1687 Isaac Newton publishes *The Mathematical Principles of Natural Philosophy*, which links astronomy with physics and puts both on firm mathematical ground.

Comet

William Herschel's
40-foot reflector

1758	Comet Halley returns as forecast by Edmond Halley, the first predicted return of a comet.
1781	William Herschel discovers Uranus, the first planet discovered since prehistoric times.
1833	A Leonid meteor shower shows that meteors come from space, not the atmosphere.
1860s	The spectroscope is invented.
1916	Albert Einstein publishes his general theory of relativity, which predicts that the Universe is expanding.
1931	Karl Jansky detects radio waves from space.
1948	Jan Oort proposes that comets come from a vast cloud orbiting far beyond Pluto.
1965	Arno Penzias and Robert Wilson discover the cosmic background radiation—the faded glow of the Big Bang explosion.
1995	First planet discovered around a star other than the Sun: orbiting 51 Pegasi.

MAPPING THE STARS

Discovering Astronomy

In ancient times, when few people could read, constellations were like heavenly storybooks. When we study the stars today, with our eyes or telescopes, we share in this age-old curiosity.

POLAR STAR TRAILS
Even though Earth is rotating, it can seem as if it is the stars that are moving. Photography captures this impression.

THE BIG DIPPER OR OXEN AND PLOUGH?
Astronomical names sometimes vary. This is because different cultures have grouped the same stars in different ways.

COMET WATCH

People have always studied the skies—a sighting of the Comet Halley was recorded in 240 BC in China. We have better technology now, but the wonder of seeing such an event—like this appearance of Comet Hale-Bopp in 1997—is the same.

DID YOU KNOW?
Under a dark sky in the countryside, the naked eye can see about 2,000 stars.

Understanding the Night Sky

Though we know Earth does the actual moving, it can still be convenient to picture the sky as a sphere turning above us. The celestial sphere has features similar to latitude, longitude, the north and south pole, and the equator.

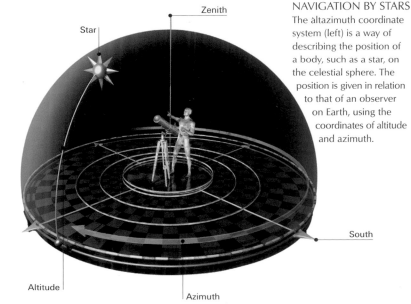

Zenith

Star

South

Altitude

Azimuth

NAVIGATION BY STARS
The altazimuth coordinate system (left) is a way of describing the position of a body, such as a star, on the celestial sphere. The position is given in relation to that of an observer on Earth, using the coordinates of altitude and azimuth.

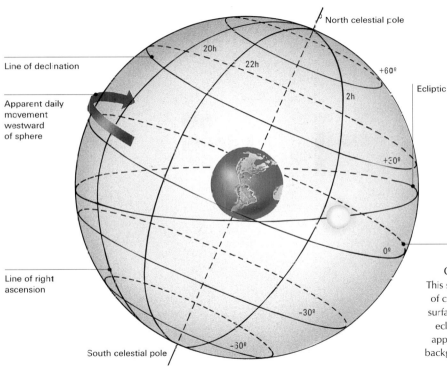

North celestial pole

Line of declination

20h

22h

+60°

Ecliptic

2h

Apparent daily
movement
westward
of sphere

+30°

Line of right
ascension

Celestial equator

0°

CELESTIAL SPHERE
This sphere and its system
of coordinates mimic the
surface of Earth. Only the
ecliptic is new—the Sun's
apparent path across the
background of the sphere.

-30°

-30°

South celestial pole

The Spinning Earth

Every hour of every day, a panorama unfolds above our heads, and all because Earth is spinning around its axis. As Earth rotates from west to east, it appears to us as if celestial objects move around us in the opposite direction.

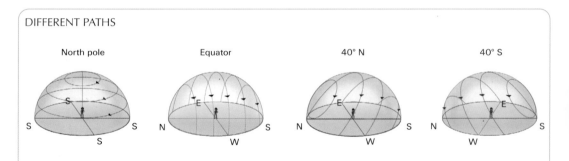

DIFFERENT PATHS

North pole | Equator | 40° N | 40° S

At the north and south pole (90°N and 90°S), the stars move parallel to the horizon. At the equator (0°), you can see all the stars of the sky, including those at the poles.

At the middle latitudes (40°N and 40°S), stars near the visible pole never set. Stars at the opposite pole are hidden from view.

A CHANGING SKY

This picture is of Orion, as seen looking south in winter from Edinburgh, Scotland. Over one hour, the stars will move from left to right.

SAME PLACE, TWO WEEKS LATER

Just as Earth's rotation makes the sky appear to move, so does Earth's movement around the Sun. At the same hour, two weeks later, Orion now appears farther west.

Measuring the Sky

When you look at the sky you will notice that stars vary in brightness. This is because some are closer to us than others, and some really are brighter than others. Astronomers describe star brightness in terms of magnitudes.

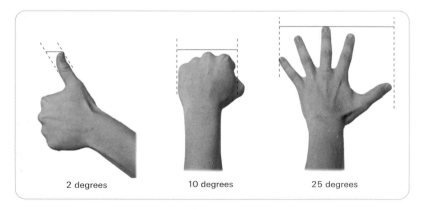

2 degrees

10 degrees

25 degrees

AT ARM'S LENGTH
Astronomers use degrees, minutes, and seconds to measure sizes in the sky. If you hold your hand at arm's length, you will see your hand covers about 25 degrees of sky. Smaller distances can be measured with the fist and thumb.

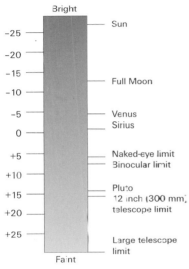

Bright

-25	Sun
-20	
-15	Full Moon
-10	
-5	Venus
0	Sirius
+5	Naked-eye limit
+10	Binocular limit
+15	Pluto 12 inch (300 mm) telescope limit
+20	
+25	Large telescope limit

Faint

APPARENT MAGNITUDE SCALE

This scale shows how objects *appear* in our sky.
The lower the number, the brighter the object;
the larger the number, the fainter the object.

APPARENT VERSUS ABSOLUTE

Absolute magnitude describes actual brightness. Orange
Algieba appears as bright as blue-white Regulus below it. But
Regulus, twice as far from Earth, is in fact much brighter.

Picturing Constellations

For as long as astronomy has existed, people have seen patterns in the stars. Some groups are so distinctive they have been known since antiquity. In the West, 88 constellations are officially recognized today, but remember, not all civilizations see the same patterns.

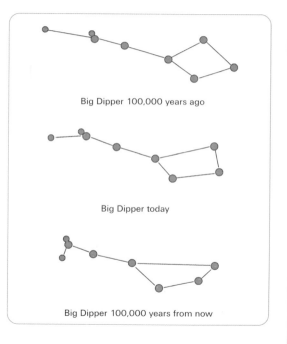

Big Dipper 100,000 years ago

Big Dipper today

Big Dipper 100,000 years from now

STARS IN MOTION

Stars are always moving, but because distances in space are so enormous, the stars in a constellation appear fixed in place. Gradually, however, their movements will make today's constellations unrecognizable.

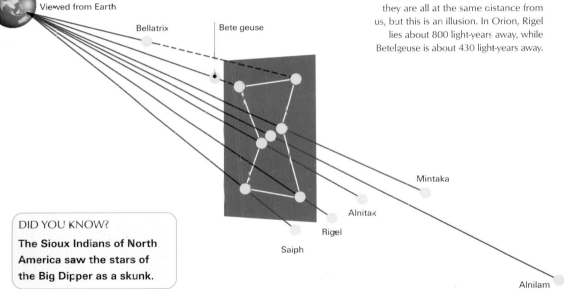

Viewed from Earth

Bellatrix

Betegeuse

The stars in a constellation look like they are all at the same distance from us, but this is an illusion. In Orion, Rigel lies about 800 light-years away, while Betelgeuse is about 430 light-years away.

Mintaka

Alnitak

Rigel

Saiph

DID YOU KNOW?
The Sioux Indians of North America saw the stars of the Big Dipper as a skunk.

Alnilam

How To Use Star Charts

The following pages show star maps for each of the four seasons for both hemispheres, beginning with spring in the Northern Hemisphere. The maps show stars, the Milky Way, star clusters, nebulas, and even galaxies, but not planets or comets as they are constantly moving.

Star magnitudes

-1	0	1	2	3	4	5

○ Open star cluster
⊕ Globular star cluster
○ Galaxy
□ Bright nebula
◇ Planetary nebula

KEY TO MAGNITUDES
Maps show star brightness using dots sized according to their magnitude.

Thin lines connect the brighter stars in major constellations.

This is the name of a star.

The symbol for open star clusters, and its name

The Milky Way in light blue

EAST

URSA MAJO

LYNX

M81 ○

UR.

Pollux

Castor

GEMINI

AURIGA

CAMELOPARDA

M35

Capella

CASSI

ORION

PERSEUS

Double Cluster

Betelgeuse

TAURUS

Algol

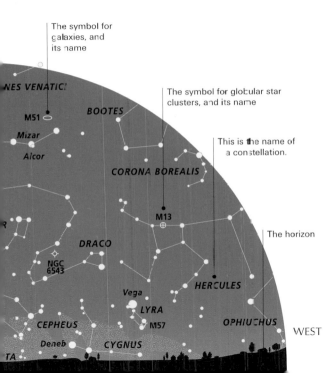

The symbol for galaxies, and its name

NES VENATIC!

M51

Mizar

Alcor

BOOTES

The symbol for globular star clusters, and its name

CORONA BOREALIS

This is the name of a constellation.

M13

DRACO

NGC
6543

The horizon

HERCULES

Vega

LYRA

M57

OPHIUCHUS

WEST

CEPHEUS

Deneb

CYGNUS

TA

STAR-HOPPING
Finding constellations can be hard. Start by identifying some of the brighter stars, such as Polaris in Ursa Minor (above), and then "star-hop."

219

Finding Your Way

To use star maps, you need to locate the four compass directions: north, south, east, and west. Each hemisphere has some bright stars that make handy starting points. Soon, you will be recognizing the brighter constellations.

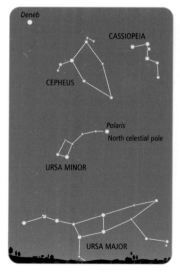

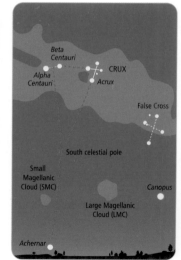

FINDING NORTH
First locate the Big Dipper in Ursa Major. Then draw an imaginary line from the end of the Dipper's bowl to the star Polaris.

THE SOUTHERN CROSS
In the Southern Hemisphere, with no bright "south star" to mark due south, the best starting place is Crux.

Northern Hemisphere Spring: Looking North

The Big Dipper in Ursa Major, the Big Bear, is upside down. The bowl points down to bright Polaris. Between them winds the snaky form of Draco the Dragon. The bright stars Pollux, Castor, Capella, and Vega line the horizon.

URSA MAJOR

LYNX

M81

URSA

Po

URSA MINOR

Pollux

Castor

GEMINI

AURIGA

CAMELOPARDAL

M35

Capella

CASSIOP

ORION

PERSEUS

Double
Cluster

Betelgeuse

TAURUS

Algol

EAST

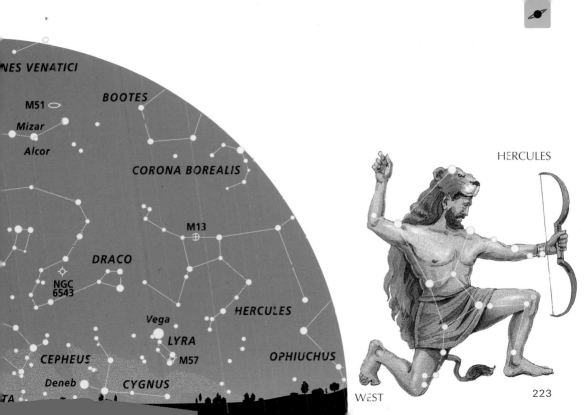

NES VENATICI

BOOTES

M51

Mizar

Alcor

CORONA BOREALIS

M13

DRACO

NGC
6543

HERCULES

Vega

LYRA

M57

CEPHEUS

OPHIUCHUS

Deneb

CYGNUS

TA

WEST

HERCULES

223

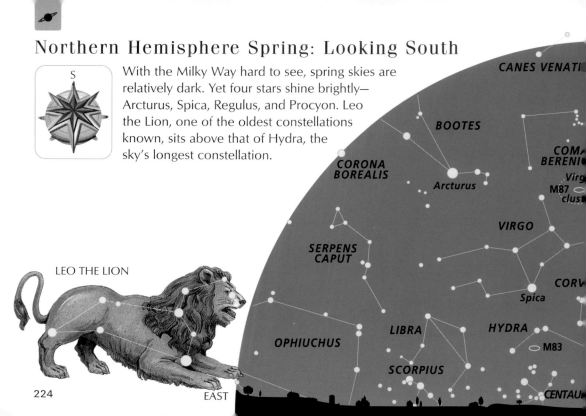

Northern Hemisphere Spring: Looking South

With the Milky Way hard to see, spring skies are relatively dark. Yet four stars shine brightly—Arcturus, Spica, Regulus, and Procyon. Leo the Lion, one of the oldest constellations known, sits above that of Hydra, the sky's longest constellation.

CANES VENATI

BOOTES

CORONA
BOREALIS

Arcturus

COMA
BERENI

Virg
M87
clust

VIRGO

SERPENS
CAPUT

LEO THE LION

CORV

Spica

LIBRA

HYDRA

OPHIUCHUS

M83

SCORPIUS

CENTAUR

EAST

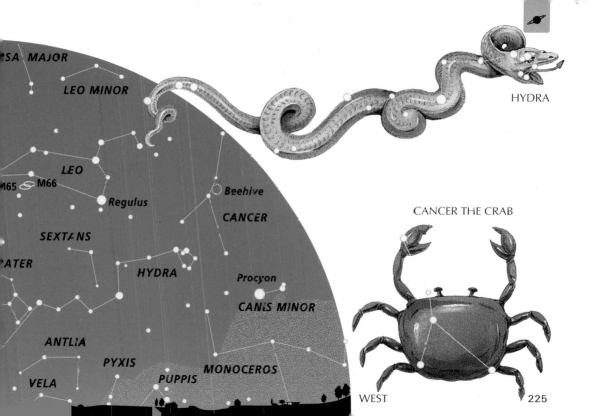

URSA MAJOR

LEO MINOR

HYDRA

LEO

M65 M66

Regulus

Beehive

CANCER

CANCER THE CRAB

SEXTANS

CRATER

HYDRA

Procyon

CANIS MINOR

ANTLIA

PYXIS

VELA

PUPPIS

MONOCEROS

WEST

Northern Hemisphere Summer: Looking North

This is the best time of the year to see the Milky Way in the Northern Hemisphere. It passes from Cassiopeia the Queen, through her husband, Cepheus, to Cygnus the Swan. The Andromeda galaxy can just be seen with the naked eye.

HERCULES

DRACO

BOOTES

M51

Mizar Alcor

URSA MINOR

Pol.

COMA BERENICES

CANES VENATICI

M81

CAM

URSA MAJOR

LEO

LYNX

LEO MINOR

CEPHEUS THE KING

WEST

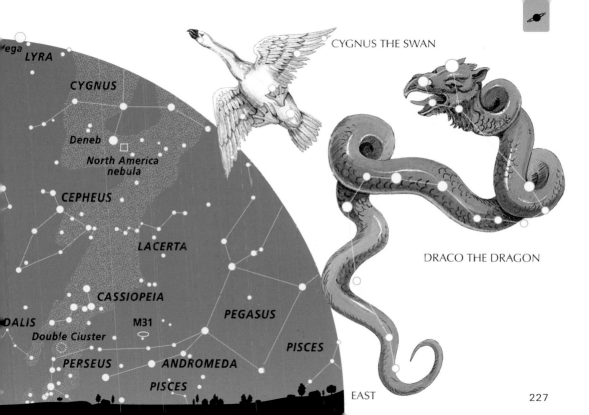

CYGNUS THE SWAN

DRACO THE DRAGON

Vega
LYRA
CYGNUS
Deneb
North America
nebula
CEPHEUS
LACERTA
CASSIOPEIA
PEGASUS
DALIS
M31
Double Cluster
PISCES
PERSEUS
ANDROMEDA
PISCES

Northern Hemisphere Summer: Looking South

The brightest part of the Milky Way parades across the southern sky on summer nights. In a dark sky, you will clearly see a rift in the Milky Way. This is caused by clouds of dust. To its right is the large but faint constellation of Ophiuchus.

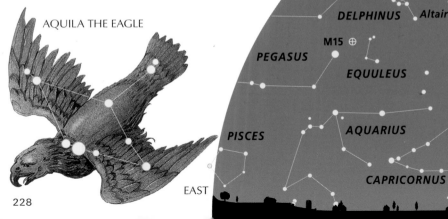

AQUILA THE EAGLE

EAST

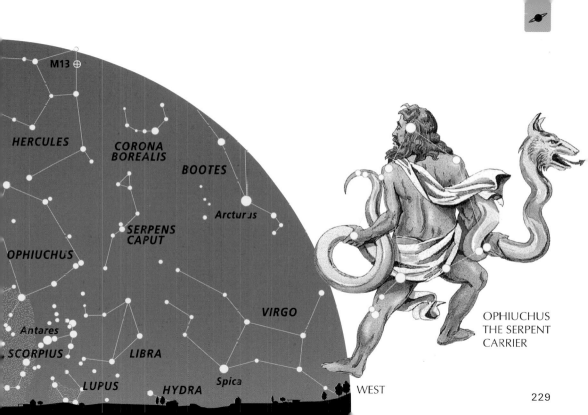

M13 ⊕

HERCULES

CORONA
BOREALIS

BOOTES

Arcturus

SERPENS
CAPUT

OPHIUCHUS

VIRGO

Antares

SCORPIUS

LIBRA

Spica

LUPUS

HYDRA

OPHIUCHUS
THE SERPENT
CARRIER

WEST

Northern Hemisphere Fall: Looking North

The small constellation of Lyra the Lyre is easy to spot at this time of the year because it has a bright star, Vega, and a distinctive shape. Auriga the Charioteer is rising in the northeast, dominated by a bright yellowish star, Capella.

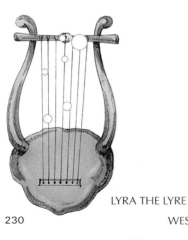

LYRA THE LYRE

LACERTA

North America nebula

Deneb

CYGNUS

CEPH

VULPECULA

LYRA

M57

M27

Vega

NGC 6543

AQUILA

DRACO   URSA MIN

M13

OPHIUCHUS

HERCULES

Alcor

Miz

WEST

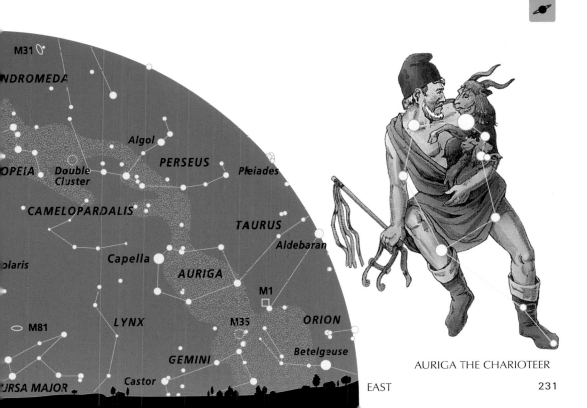

M31

NDROMEDA

M31

Algol

OPEIA   Double
         Cluster

PERSEUS

Pleiades

CAMELOPARDALIS

TAURUS

Aldebaran

olaris

Capella

AURIGA

M1

M81

LYNX

M35

ORION

Betelgeuse

GEMINI

URSA MAJOR

Castor

AURIGA THE CHARIOTEER

Northern Hemisphere Fall: Looking South

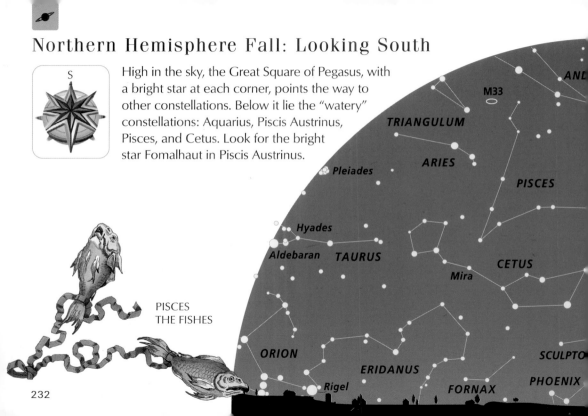

S

High in the sky, the Great Square of Pegasus, with a bright star at each corner, points the way to other constellations. Below it lie the "watery" constellations: Aquarius, Piscis Austrinus, Pisces, and Cetus. Look for the bright star Fomalhaut in Piscis Austrinus.

M33

AND

TRIANGULUM

ARIES

PISCES

Pleiades

Hyades

Aldebaran TAURUS

CETUS

Mira

PISCES
THE FISHES

ORION

SCULPTO

ERIDANUS

PHOENIX

Rigel FORNAX

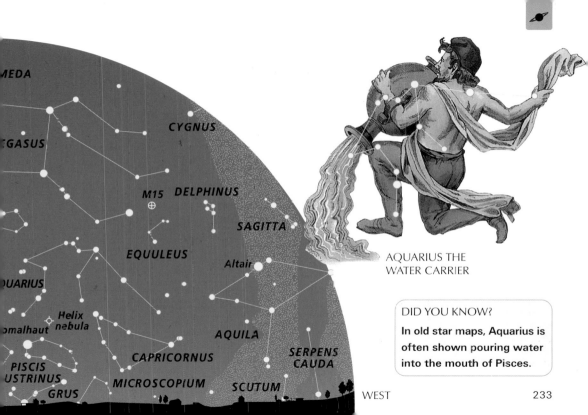

MEDA

CYGNUS

PEGASUS

M15 ⊕ DELPHINUS

SAGITTA

EQUULEUS

Altair

AQUARIUS

Helix
nebula

AQUILA

Fomalhaut

SERPENS
CAUDA

PISCIS
AUSTRINUS

CAPRICORNUS

MICROSCOPIUM

SCUTUM

GRUS

AQUARIUS THE
WATER CARRIER

DID YOU KNOW?

In old star maps, Aquarius is
often shown pouring water
into the mouth of Pisces.

WEST

Northern Hemisphere Winter: Looking North

N

Many people can recognize the Big Dipper (known as the Plough in Europe). It is made from the seven bright stars that mark out the back and tail of Ursa Major. The star Algol in Perseus varies in brightness every three days for a few hours.

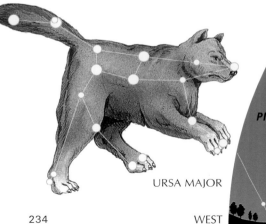

URSA MAJOR

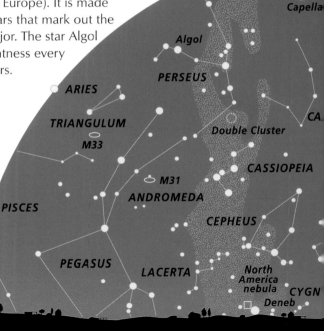

Capella

Algol

PERSEUS

ARIES

TRIANGULUM

M33

CA.

Double Cluster

M31

CASSIOPEIA

ANDROMEDA

PISCES

CEPHEUS

PEGASUS

LACERTA

North America nebula

CYGN

Deneb

WEST

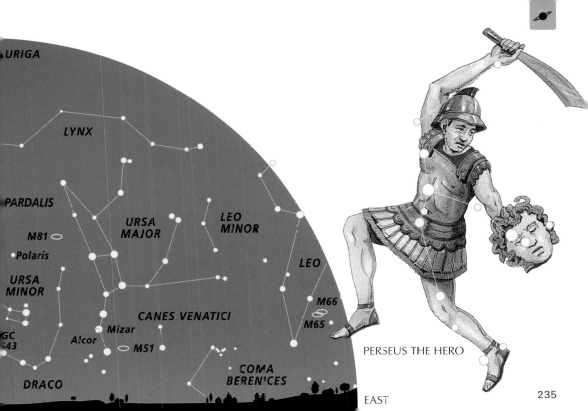

AURIGA

LYNX

PARDALIS

M81

Polaris

URSA MINOR

URSA MAJOR

LEO MINOR

LEO

NGC 243

Alcor   Mizar

CANES VENATICI

M51

M66

M65

COMA BERENICES

DRACO

PERSEUS THE HERO

EAST

235

Northern Hemisphere Winter: Looking South

S

Orion the Hunter stands high above the horizon. From the middle of his belt hangs his sword, which contains the Orion nebula (M42), just visible to the eye. Orion is driving back Taurus the Bull, which has the lovely Pleiades star cluster on its shoulder.

TAURUS THE BULL

Castor

Pollux

M3?

GEMINI

Beehive

CANCER

CANIS MINOR

LEO

Regulus

Procyon

MONOCEROS

HYDRA

Sirius

CANIS MAJOR

M41

SEXTANS

Adhar

PYXIS

PUPPIS

ANTLIA

EAST

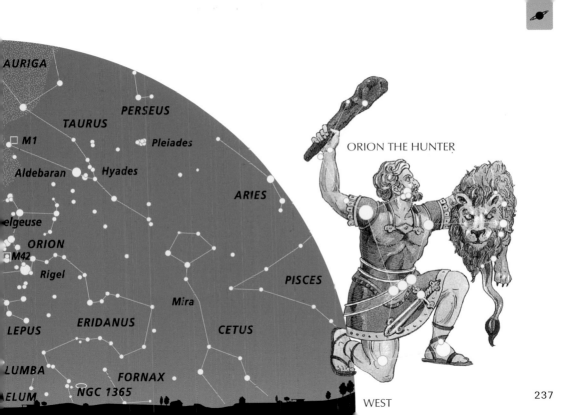

AURIGA

PERSEUS

TAURUS

M1

Pleiades

Aldebaran   Hyades

ARIES

elgeuse

ORION

M42

Rigel

PISCES

Mira

LEPUS   ERIDANUS   CETUS

LUMBA   FORNAX

ELUM   NGC 1365

ORION THE HUNTER

WEST

237

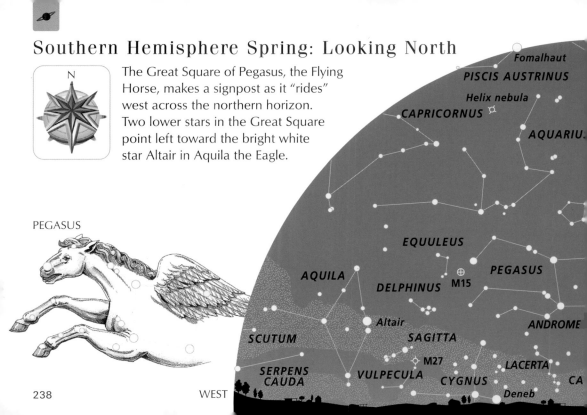

Southern Hemisphere Spring: Looking North

The Great Square of Pegasus, the Flying Horse, makes a signpost as it "rides" west across the northern horizon. Two lower stars in the Great Square point left toward the bright white star Altair in Aquila the Eagle.

PEGASUS

Fomalhaut
PISCIS AUSTRINUS
Helix nebula
CAPRICORNUS
AQUARIU.
EQUULEUS
PEGASUS
AQUILA
DELPHINUS M15
ANDROME
Altair
SAGITTA
SCUTUM
LACERTA
SERPENS CAUDA
VULPECULA
CYGNUS
CA
M27
Deneb

238

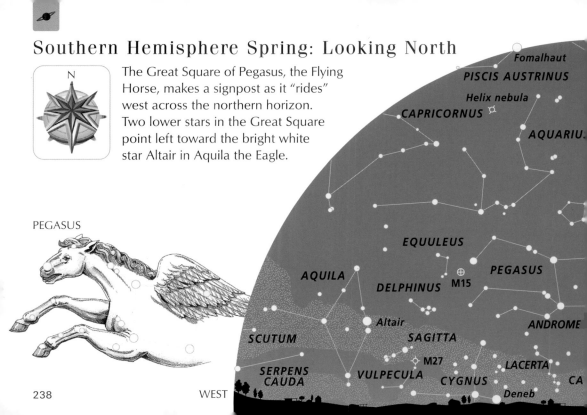

WEST

CULPTOR

CETUS

ERIDANUS

PISCES

Mira

ARIES

TAURUS

M33

Pleiades

M31

TRIANGULUM

Hyades   Aldebaran

ORION

PEIA   PERSEUS   Algol

DID YOU KNOW?

In the neck of Cetus the Sea Monster lies the star Mira. It brightens and fades from view every 11 months.

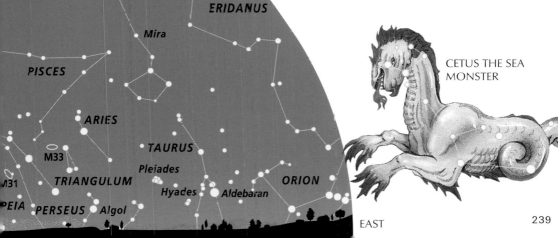

CETUS THE SEA MONSTER

239

Southern Hemisphere Spring: Looking South

Two bright stars catch your eye at this time of the year: Achernar in the winding Eridanus the River; and Canopus in Carina the Keel. Canopus is the sky's second-brightest star, after Sirius. Celestial birds include Phoenix the Firebird, Grus, and Tucana.

S

SCULPTO

PHOENIX

FORNAX

ERIDANUS

NGC 1365

Achernar

HOROLOGIUM

47

RETICULUM   SMC
HYDRU.

CAELUM   DORADO   Tarantula
nebula

LEPUS   PICTOR   MENS

LMC
CHAMAELEC

Rigel   Canopus   VOLANS

M42   COLUMBA   CARINA

TUCANA THE TOUCAN

ORION CANIS MAJOR   PUPPIS

Adhara   VELA

EAST

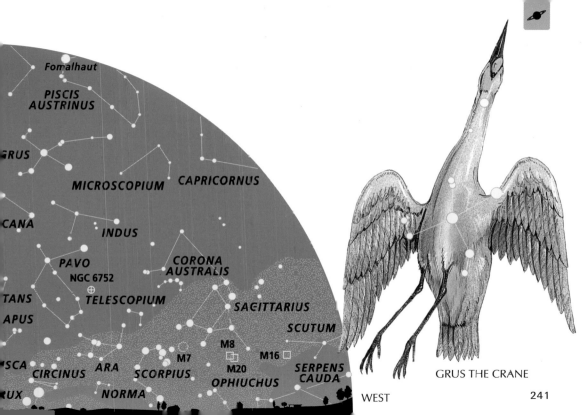

Fomalhaut

PISCIS
AUSTRINUS

GRUS

MICROSCOPIUM

CAPRICORNUS

CANA

INDUS

PAVO
NGC 6752

CORONA
AUSTRALIS

TANS

TELESCOPIUM

SAGITTARIUS

APUS

SCUTUM

M8

M16

SCA

CIRCINUS   ARA   SCORPIUS

M7

M20

OPHIUCHUS

SERPENS
CAUDA

UX

NORMA

GRUS THE CRANE

Southern Hemisphere Summer: Looking North

N

If you look toward Orion, you will notice how the colors of Betelgeuse (a cool, red star) and Rigel (a hot, blue one) differ. Just to the east lies Sirius, the brightest star in the sky. Lower down lie the two bright stars Castor and Pollux in Gemini.

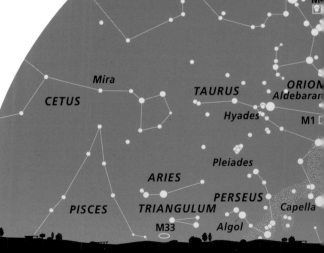

LEPUS

ERIDANUS

Rigel

M-

Mira

CETUS

TAURUS

ORION

Aldebaran

Hyades

M1

Pleiades

ARIES

PERSEUS

Capella

PISCES

TRIANGULUM

Algol

M33

GEMINI
THE TWINS

WEST

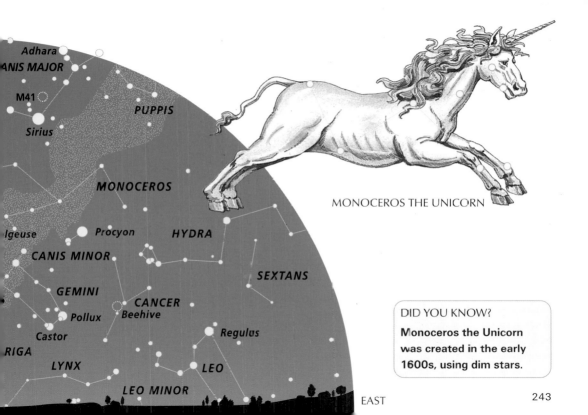

Adhara

CANIS MAJOR

M41

Sirius

PUPPIS

MONOCEROS

Igeuse

Procyon

HYDRA

CANIS MINOR

SEXTANS

GEMINI

CANCER

Pollux   Beehive

Castor

Regulus

RIGA

LYNX

LEO

LEO MINOR

MONOCEROS THE UNICORN

DID YOU KNOW?

**Monoceros the Unicorn
was created in the early
1600s, using dim stars.**

Southern Hemisphere Summer: Looking South

High in the south, the bright star Canopus marks the rudder of Carina the Keel. Carina is part of the mythical ship *Argo*. Other parts of the *Argo* include Vela the Sails, Puppis the Stern, and Pyxis the Compass, all in the bright summer Milky Way.

Adhara
CANIS MAJOR
COLUM
PUPPIS
Cano
PYXIS
PICTO
CARINA
ANTLIA
False Cross
VOLANS
VELA
HYDRA
Eta Carinae nebula
CHAMAELEC
MUSCA
CENTAURUS Acrux
Coalsack nebu
CRATER
CRUX Jewel Box
CORVUS
Beta Cen
AP
Omega Centauri ⊕
Alpha Cen
CIRCIN

COLUMBA THE DOVE

EAST

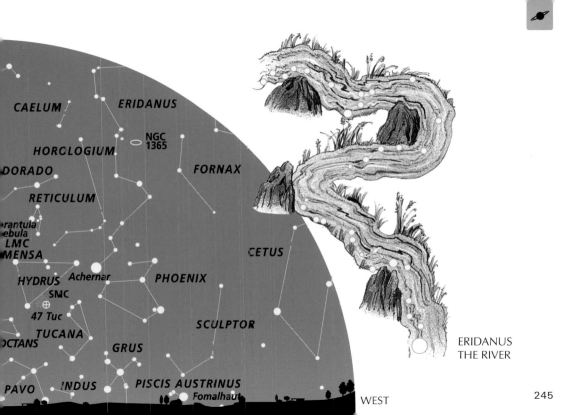

CAELUM

ERIDANUS

HOROLOGIUM

NGC
1365

DORADO

FORNAX

RETICULUM

Tarantula
Nebula
LMC
MENSA

CETUS

HYDRUS   Achernar   PHOENIX

SMC
47 Tuc

OCTANS

TUCANA

SCULPTOR

GRUS

PAVO   INDUS   PISCIS AUSTRINUS
Fomalhaut

ERIDANUS
THE RIVER

Southern Hemisphere Fall: Looking North

Four bright stars make easy jumping-off points for finding constellations right now: Procyon in Canis Minor the Little Dog; Regulus, which marks the heart of Leo the Lion; Spica in Virgo the Maiden; and Arcturus in the constellation Boötes the Herdsman.

VIRGO THE MAIDEN

WEST

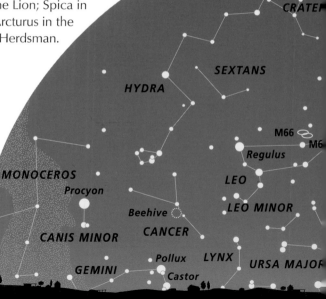

CRATER

SEXTANS

HYDRA

M66

M6

Regulus

LEO

MONOCEROS

Procyon

LEO MINOR

Beehive

CANCER

CANIS MINOR

Pollux

LYNX

URSA MAJOR

GEMINI

Castor

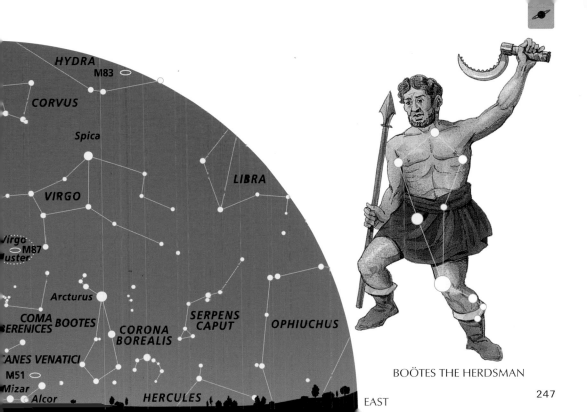

HYDRA
M83

CORVUS

Spica

LIBRA

VIRGO

Virgo
M87
Cluster

Arcturus

COMA
BERENICES BOOTES
CORONA
BOREALIS
SERPENS
CAPUT
OPHIUCHUS

CANES VENATICI

M51

Mizar
Alcor

HERCULES

EAST

BOÖTES THE HERDSMAN

Southern Hemisphere Fall: Looking South

If you have a pair of binoculars or a telescope, this is a great time to have a look at the Milky Way stars. The band runs from the dazzling star Sirius in the west to Antares in the southeast. High in the sky, Centaurus hops over Crux, the Southern Cross.

SCORPIUS THE SCORPION

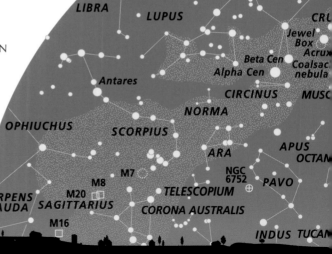

HYDRA

CENT

Omega
Centauri

LIBRA

LUPUS

CRU

Jewel
Box

Acrux

Beta Cen   Coalsac

Alpha Cen   nebula

CIRCINUS   MUSC

Antares

NORMA

OPHIUCHUS   SCORPIUS

ARA   APUS

OCTAN

M7

NGC
6752   PAVO

M8

M20   TELESCOPIUM

SERPENS
CAUDA   SAGITTARIUS   CORONA AUSTRALIS

M16

INDUS TUCAN

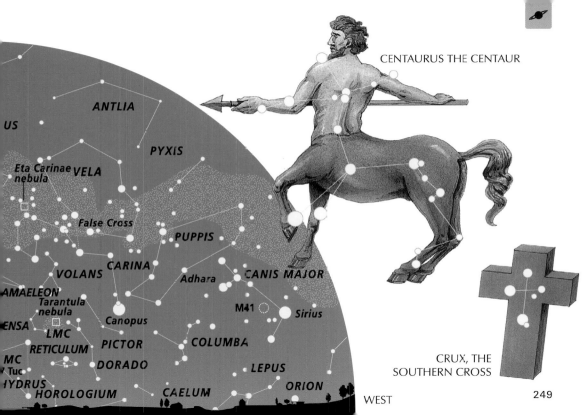

CENTAURUS THE CENTAUR

ANTLIA

US

PYXIS

Eta Carinae
nebula

VELA

False Cross

PUPPIS

CARINA

VOLANS

Adhara

CANIS MAJOR

AMAELEON

Tarantula
nebula

Canopus

M41

Sirius

ENSA

LMC

RETICULUM

PICTOR

COLUMBA

MC

Tuc

DORADO

LEPUS

HYDRUS

HOROLOGIUM

CAELUM

ORION

CRUX, THE
SOUTHERN CROSS

Southern Hemisphere Winter: Looking North

N

High overhead lies Sagittarius, marking the center of our galaxy. Leading him across the sky are the bright stars of Scorpius and the fainter stars of Libra. Look out also for bright Arcturus, white Vega in Lyra the Lyre, and Altair in Aquila the Eagle.

Antares

SCORPIUS

LIBRA

OPHIUCHUS

SAGITTARIUS
THE ARCHER

SERPENS
CAPUT

HERCUL

Spica

CORONA
Arcturus BOREALIS

M13 ⊕

BOOTES

VIRGO

DRACO

WEST   COMA BERENICES

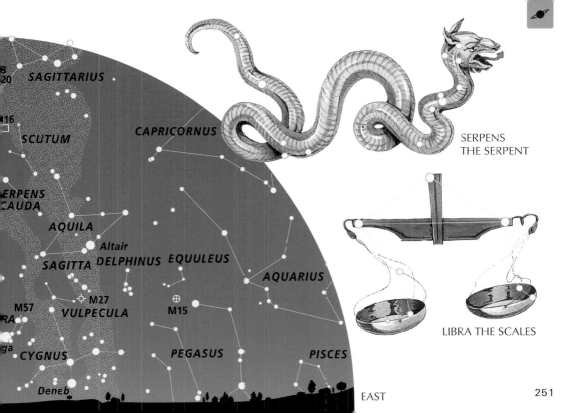

SAGITTARIUS

SCUTUM

CAPRICORNUS

SERPENS
THE SERPENT

SERPENS
CAUDA

AQUILA

Altair

SAGITTA DELPHINUS EQUULEUS

AQUARIUS

M57

M27

⊕
M15

VULPECULA

LIBRA THE SCALES

CYGNUS

PEGASUS

PISCES

Deneb

Southern Hemisphere Winter: Looking South

Now the bright southern Milky Way swings low into the southwest. Look for Crux and the dark Coalsack nebula, a large patch of interstellar dust. Curving above Crux is Centaurus, half-man, half-horse. Alpha and Beta Centauri mark its forelegs.

CORONA AUSTRALIS

EAST

SAGITTARIUS

CORONA AUSTRALIS

CAPRICORNUS

TELESCOPIUM

MICROSCOPIUM

INDUS

PAVO

PISCIS AUSTRINUS

Helix nebula

GRUS

OCTANS

Fomalhaut

AQUARIUS

TUCANA

47 Tuc

SMC HYDRUS

SCULPTOR

PHOENIX

MENSA

Achernar LMC

ERIDANUS

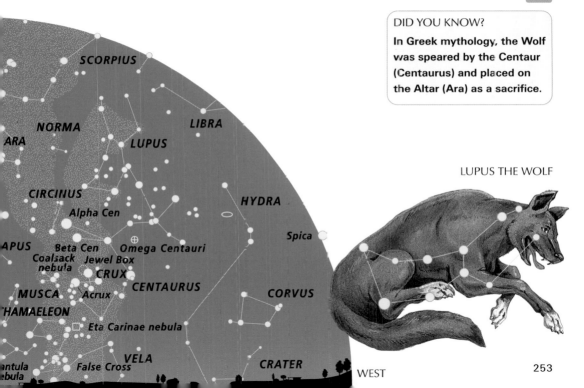

SCORPIUS

LIBRA

NORMA

ARA

LUPUS

CIRCINUS

HYDRA

Alpha Cen

LUPUS THE WOLF

Spica

APUS

Beta Cen   Omega Centauri

Coalsack   Jewel Box
nebula

CRUX

MUSCA   Acrux

CENTAURUS

CORVUS

CHAMAELEON

Eta Carinae nebula

VELA

antula
ebula

False Cross

CRATER

WEST

253

OBSERVING THE SKY

Using your Eyes

Stargazing is easy, and you can start with nothing more than an uninterrupted view of the sky, some comfortable clothing, and your own pair of eyes. The unaided eye is capable of finding constellations, and observing meteor showers and comets. As well, five of the planets can be seen with the naked eye—Mercury, Venus, Mars, Jupiter, and Saturn.

GALAXY SPOTTING
Two medium-sized members of the Local Group can be seen in southern skies: the Small (left) and Large Magellanic clouds (right).

BRIGHT PLANETS
This evening sky shows the crescent Moon with two clearly visible planets—Venus above it, near the top of the photo, and Jupiter below.

City Lights

Cities are not the best places for stargazing because of the amount of light that is always around. For best results, try to find a dark place such as a park, away from the city.

THE WORLD LIT UP
In this NASA image of Earth, North America and Europe have the greatest concentration of city lights.

CITY LIGHTS
In densely populated areas, few locations can be said to be truly dark.

STARRY NIGHTS
To really see the bright lights, you need to escape the city and head to wide, open country areas.

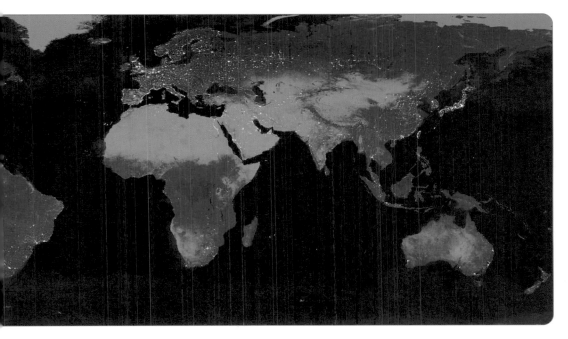

Binoculars

Using binoculars is an easy way to view a wide range of celestial objects. Most are lightweight, portable, and require no setting-up—you can just pick them up and away you go.

USING BINOCULARS
Binoculars are essentially two low-powered telescopes joined together, so that you can look through them with both eyes instead of just one. They are easy to use, which is important for kids.

HOLD STEADY
A camera tripod will hold binoculars steady. Alternatively, steady yourself against a wall.

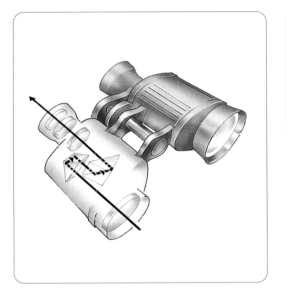

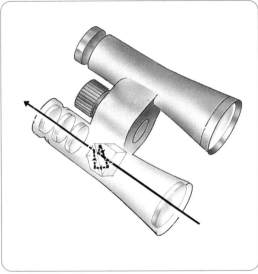

PORRO PRISM
This is the style familiar to most skywatchers. The shape of the binoculars results from the prism arrangement which produces an image that is upright and the right way round.

ROOF PRISM
Roof prism binoculars are usually smaller than porro-prism models. They use a different arrangement of prisms to achieve the same result.

Choosing the Right Pair

The performance of binoculars depends on two things: the diameter of the lenses at the front; and the magnification of the eyepieces. A code is used to indicate these specifications. For example, 7 x 50 means a magnification of 7, and a diameter of 50 mm.

7 x 50   9 x 63

A BALANCING ACT
A power between 7 and 10 is ideal for most users. Too high and the image will wobble. Increasing the lens diameter lets more light in, but the bigger it is, the heavier they will be.

DIFFERING VIEWS
These views of the Pleiades show the relationship between field of view and magnification. The greater magnification (right) offers more detail but the field of view is reduced.

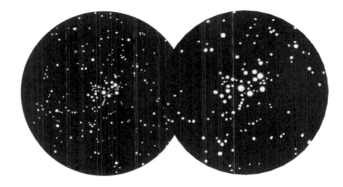

BUYING WISELY
Try out a few binoculars before buying. Check how they fit in your hands and on the eyes.

Telescopes

A telescope can reveal small craters on the Moon, Jupiter's clouds, Saturn's rings, and also star clusters, nebulas, and galaxies that are too faint for binoculars. For the keen astronomer, they can be highly rewarding. Various types exist, but the most important feature is the size of the mirror or lens—the bigger, the better.

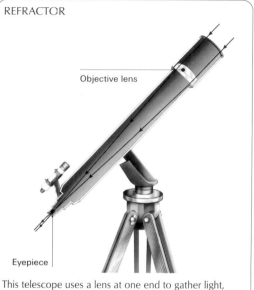

REFRACTOR

Objective lens

Eyepiece

This telescope uses a lens at one end to gather light, which is focused to an eyepiece at the other end.

DID YOU KNOW?

Generally, a reflecting telescope will provide larger optics than a refractor of the same size. The result is more light coming into your eye.

NEWTONIAN REFLECTOR

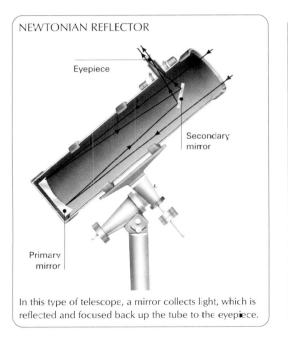

Eyepiece

Secondary mirror

Primary mirror

In this type of telescope, a mirror collects light, which is reflected and focused back up the tube to the eyepiece.

CATADIOPTRIC

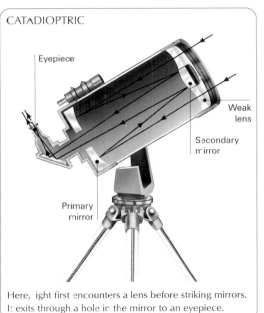

Eyepiece

Weak lens

Secondary mirror

Primary mirror

Here, light first encounters a lens before striking mirrors. It exits through a hole in the mirror to an eyepiece.

Telescope Mounts

Stability is essential when using a telescope.
There is nothing worse than a flimsy mount or
spindly tripod that produces images that never
stop bouncing! There are two main telescope
mounts: altazimuth and equatorial.

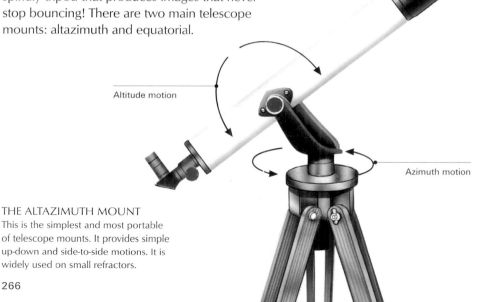

Altitude motion

Azimuth motion

THE ALTAZIMUTH MOUNT
This is the simplest and most portable
of telescope mounts. It provides simple
up-down and side-to-side motions. It is
widely used on small refractors.

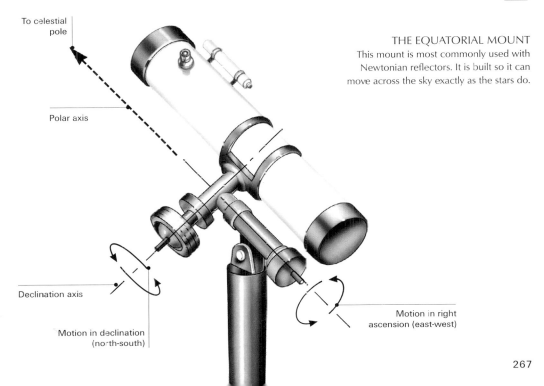

To celestial pole

Polar axis

Declination axis

Motion in declination
(north-south)

THE EQUATORIAL MOUNT
This mount is most commonly used with
Newtonian reflectors. It is built so it can
move across the sky exactly as the stars do.

Motion in right
ascension (east-west)

Telescope Accessories

Whenever you look through a telescope, you look through an eyepiece, and these are often the first accessories telescope owners buy. It doesn't stop there, however, with everything from solar filters to computer controls on offer.

DID YOU KNOW?
With a motor drive, you can view an object without even touching the telescope.

EYEPIECES
A telescope's main lens or mirror gathers the incoming light and focuses it into an image, but it is the eyepiece that magnifies the image. To change magnifications, just change the eyepiece.

Star Photography

Capturing images of the stars doesn't have to mean expensive equipment. All you need is some fast film, a camera, and a tripod.

PIN-POINT SKY
Bolting the camera to a motor-driven equatorial mount helps capture fainter stars by tracking the sky.

AIMING AT THE SKY
Cameras with mechanical shutters can be more reliable than electronic models.

Viewing the Sun

The Sun was probably the very first object to be studied. Today, the fascination remains—as does the need to take great care when observing the Sun. Never view it through binoculars or a telescope without a filter, as it takes only a split second to be blinded by unfiltered sunlight.

SPOTS ON THE SUN
The movement of sunspots reveals that the Sun rotates about once a month.

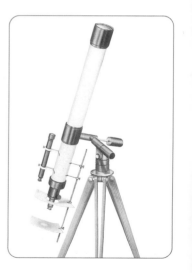

SAFE VIEWING
This telescope is equipped with a Sun projection screen. The Sun's image is focused onto the lower screen, while the upper screen blocks other sunlight.

SOLAR FILTERS
The best way to view the Sun safely is to fit a solar filter over the telescope's front lens.

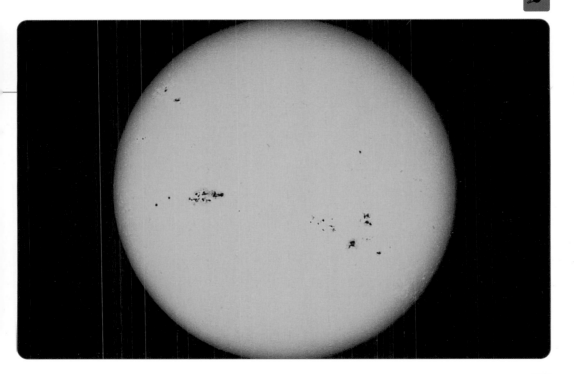

PARTIAL ECLIPSE
A total eclipse is a rare event, though a partial one still makes striking viewing.

OBSERVING A PARTIAL ECLIPSE
However tempting it may be, never look directly at a partially eclipsed Sun.

THE SUN'S CORONA
The normally invisible corona streams out at the moment of total eclipse.

TOTAL SOLAR ECLIPSE
As a total eclipse occurs, the last burst of sunlight resembles a diamond ring.

Solar Eclipses

An eclipse of the Sun can be partial, annular, or total. This spectacular event occurs somewhere on Earth on average once every 19 months.

The Moon

Earth is unique among the inner planets in having a large natural satellite—the Moon. Due to bombardment by asteroids and meteorites, its surface is a maze of lava-filled basins (called "seas"), craters, and bright streaks known as rays. These have been caused by relatively recent impacts.

AN ANCIENT FACE
The face of the Moon has appeared essentially unchanged for more than half Earth's history. Images from space reveal fascinating details of its surface.

MOONSHINE
The Moon has been Earth's companion for over 4 billion years. Some believe it was once part of our planet.

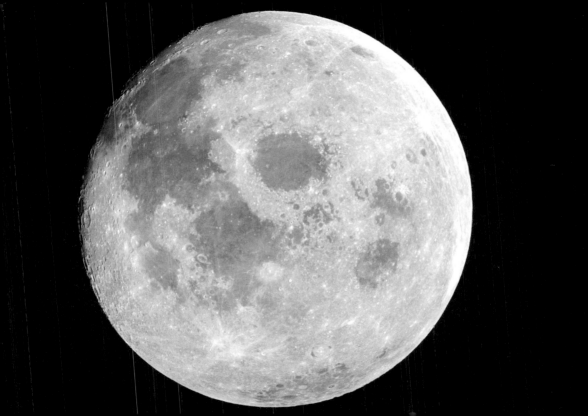

Auroras

Auroras take many forms. Pulsating arcs, shimmering curtains of color, and flickering shafts of light are the most common and striking. The displays usually appear light green, with reddish tints indicating more energetic activity. Auroras are most common at high latitudes, especially the polar regions.

DIVINE MESSAGES
Auroras were once believed to be divine messages. And, indeed, the ghostly arcs of shimmering light do seem out of this world.

CURTAINS OF LIGHT
Auroras are most visible around Earth's magnetic poles. In the north they are called aurora borealis and in the south, aurora australis. Shown here is an aurora australis as seen from space.

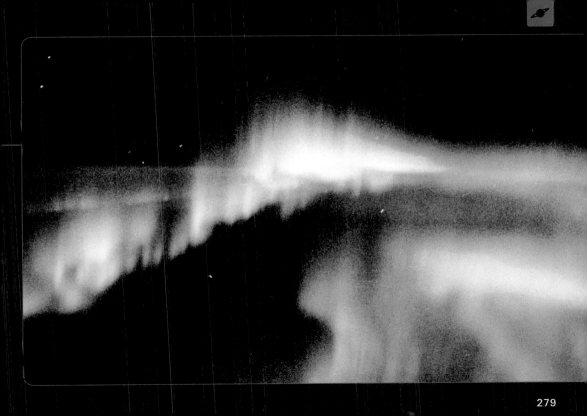

Comets and Meteors

A bright comet is a wonderful sight, be it with the unaided eye, in binoculars, or through a telescope. The only drawback is their unpredictability! Meteors are much more common and, at certain times of the year, appear in large numbers.

METEOR SHOWER
Most meteors are lone streaks of light. A meteor shower occurs when Earth passes near some cometary debris.

COMET HYAKUTAKE
The appearance of a bright comet is usually a surprise. This comet arrived unheralded in 1996 and stayed for a few weeks.

COMET HALE-BOPP
Everyone can enjoy the sight of a comet. Comet Hale-Bopp was a bright naked-eye comet with two distinct tails.

Viewing the Planets

Not every planet can be observed easily, but Venus, Mars, Jupiter, and Saturn all make excellent subjects. Venus, in particular, shines out, even among bright city lights. As well, some moons and rings of planets are visible.

POLAR CAPS OF MARS
Seeing detail on Mars can be difficult but the ice caps often stand out. The white southern cap is visible here.

A BRIGHT NEIGHBOR
When seen through a telescope, Venus can be blindingly bright. In these images, it peeps out from behind the crescent Moon.

MARS ON THE MOVE
Mars can appear at any place in the sky near the ecliptic (the Sun's apparent path in the sky). Here, it glows a dull orange (right), rivaled by the star Antares (center).

The Outer Planets

To the naked eye, Jupiter and Saturn look like bright "stars." Telescopes, however, can reveal much more: red spots and colored bands on Jupiter; the magnificent rings around Saturn; and the four large moons of Jupiter.

CHANGING ASPECTS
Because Saturn's axis tilts 27 degrees to its orbit, our view of the rings is constantly changing.

WHAT GALILEO SAW
A small telescope lets you share in Galileo's discovery—Jupiter's largest moons.

OBSERVING JUPITER

As Jupiter makes its way slowly along the ecliptic, it passes
various clusters, nebulas, and bright stars. Here it lies near
the lovely Beehive Cluster in Cancer.

Alcor and Mizar and the Pleiades

Observing Alcor and Mizar may reveal three, not two, stars. The Pleiades is the most famous open star cluster in the sky. Those with sharp eyes might be able to recognize nine stars.

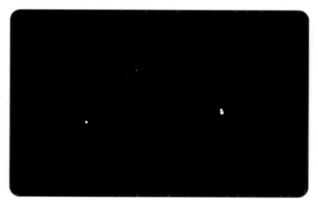

ALCOR AND MIZAR
The Big Dipper's handle consists of Alcor (left) and Mizar (right). Mizar is actually a double star—you can just see its companion here.

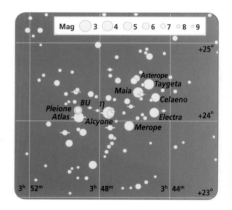

THE PLEIADES REGION
With over 500 stars, it can be a good idea to use a star chart when viewing the Pleiades.

Double Cluster and Jewel Box Cluster

These two sparkling star clusters, one in each hemisphere, are good targets for binoculars. If you are lucky enough to be viewing the Jewel Box with a telescope, just remember that too much magnification will "uncluster" the cluster.

The Jewel Box looks like a small star, tucked under Beta Crucis, the bright lefthand point of the Southern Cross. Near it is the dark Coalsack nebula.

DOUBLE STAR CLUSTER
This magnificent pair of clusters has long been a favorite target for Northern Hemisphere skywatchers. The clusters are NGC 884 (left) and NGC 869 (right), though they are not actually related.

Orion Nebula and Lagoon Nebula

These two diffuse nebulas—one for summer and one for winter—are both visible to the naked eye. They make superb targets for binoculars or a small telescope. Both nebulas are "star factories"—the place where new stars are formed.

LAGOON NEBULA

This giant cloud of dust and hydrogen gas may produce thousands of stars. Images from orbiting spacecraft reveal a place of intense energy and activity.

ORION NEBULA

Even from a city, the Orion Nebula is visible to the naked eye as a bright fuzzy patch. It is located in Orion's sword, just south of his belt.

Hercules Cluster and Omega Centauri

The Hercules globular cluster consists of hundreds of thousands of stars tightly squeezed into a relatively small space. In contrast, Omega Centauri has 10 times the mass of other large globulars. It is perhaps the sky's finest globular cluster.

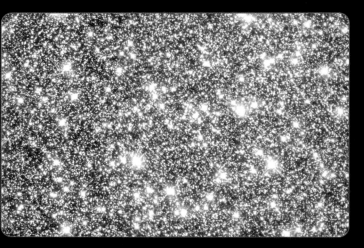

HERCULES CLUSTER
This spectacular globular cluster in the northern skies has up to 1 million stars. When you view it, you are looking 23,000 years into the past.

A MASS OF STARS
Omega Centauri is so bright that it was originally mistaken for a star. Pictures such as this one show just how densely packed with stars it is.

Andromeda Galaxy and Pinwheel Galaxy

These two Northern Hemisphere galaxies are quite distinct: the Andromeda Galaxy is very bright and large, while the Pinwheel is delicate and difficult to locate. Both are significant members of our Local Group of galaxies.

2.9 MILLION LIGHT-YEARS AWAY
The Andromeda Galaxy is the most distant object visible to the naked eye—though not with this much detail!

PINWHEEL GALAXY
Although bright, the Pinwheel Galaxy is hard to see, it appears face-on and is spread over a large patch of sky.

Ring Nebula and Trifid Nebula

The skies of the Northern Hemisphere are graced by the striking Ring Nebula, perhaps the most popular planetary nebula for viewing. In the Southern Hemisphere, keen skywatchers may be able to pick out the beautiful but diffuse Trifid Nebula.

DID YOU KNOW?

Our eyes are not sensitive enough to see a nebula's pretty colors—to us, they all look gray-green.

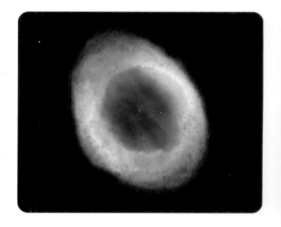

REMNANT OF A STAR
The Ring Nebula is the exhaled shell of a dying star, shown in this Hubble Space Telescope image in brilliant detail. Despite its name, the nebula is more barrel-shaped, not ring-shaped.

TRIFID NEBULA
Long-exposure photography reveals three lanes of dark cloud dividing this nebula. This feature has earned it the name trifid, meaning to split into three.

Eagle Nebula

The Eagle is truly amazing: a combination of nebula and star cluster—the stars having been born within the nebula itself. Large telescopes reveal the fantastic pillars of dark interstellar dust at the nebula's center.

NURSERIES FOR STARS
Deep within these columns of gas several light-years in length, stars are forming. Astronomers believe that the Universe contains billions of these stellar nurseries.

INSIDE THE EAGLE NEBULA
Images taken from telescopes on Earth show hot, young stars inside the nebula. Our Sun and planets formed out of a similar nebula.

Crab Nebula and Dumbbell Nebula

In 1054, Chinese astronomers witnessed a massive explosion, known as a supernova. Nearly a thousand years later, when we look at the Crab Nebula, we are looking at the shattered remains of that star spread out like a web. Similarly, the Dumbbell Nebula is a cloud of gas thrown off by an old star.

CRAB NEBULA
The jettisoned remains of this star are found in the constellation Taurus. To see it in detail, a fairly large telescope is needed.

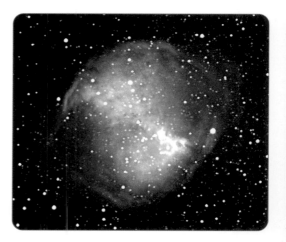

DUMBBELL NEBULA
This is one of the largest planetary nebulas. Softly glowing and green, it is about 1,000 light-years away.

Whirlpool Galaxy and M81

The Whirlpool Galaxy was the first galaxy in which spiral arms were clearly identified. It has a slightly frenzied appearance, unlike the remarkably symmetrical shape of M81. This spiral galaxy is one of several galaxies that form a small galaxy cluster called the M81 group.

M81 SPIRAL GALAXY
This galaxy is a real skywatcher's delight—a bright galaxy visible in binoculars even from a city. The arms are faint, but the central core is bright.

WHIRLPOOL GALAXY
This striking face-on spiral galaxy has a distinctive hurricane-like appearance—hence its name. At the end of one of the spiral arms is a smaller galaxy called NGC 5195. The two are engaged in a gravitational tug-of-war.

Large Magellanic Cloud and Great Barred Spiral

The Milky Way's nearest galactic neighbor is the Large Magellanic Cloud (LMC). It has hundreds of millions of stars, and is rich in star formation. Also in the Southern Hemisphere is the Great Barred Spiral. Found in the constellation Fornax, it has a prominent central region and dramatic scythe-like arms.

A SUPERGIANT GALAXY
The Great Barred Spiral, seen here in close up, is truly big. It has a diameter of about 200,000 light-years.

LARGE MAGELLANIC CLOUD
Lying 160,000 light-years from the Milky Way, this "cloud" of stars is an entire galaxy, in orbit around our own.

Glossary

absolute magnitude Brightness of a star at a standard distance of 10 parsecs (32.6 light-years) from Earth.

active galaxy A galaxy that emits an excess of radiation, perhaps from gas falling into a central black hole.

aperture Diameter of the main light-collecting optics on a telescope. Also, diameter of binocular lens.

apparent magnitude The visible brightness of a star or celestial body as seen from Earth.

asteroid Metallic or stony object (under 600 miles (1,000 km) across) orbiting the Sun. Also a minor planet.

asteroid belt Reservoir of asteroids orbiting the Sun between the orbits of Mars and Jupiter.

atmosphere A layer of gases encircling a planet, star, or moon, and held by the body's gravity.

aurora Curtains and arcs of light appearing in the sky over middle and high latitudes. They are caused by particles from the Sun hitting Earth's atmosphere and causing some of its gases to glow.

axis The imaginary line through a planet, star, moon, or galaxy around which it rotates; also, a shaft around which a telescope mounting pivots.

Big Bang The theoretical eruption of a very hot lump of energy about 14 billion years ago that marked the birth of the Universe.

binary star Two stars linked by mutual gravity and revolving around a common center of mass.

black hole An object so dense that no light or other radiation can escape its gravitational pull.

celestial sphere The imaginary sphere around Earth upon which stars and other objects appear to lie.

comet A small body composed of ice and dust that orbits the Sun.

constellation Patterns of stars that divide the sky into sections. There are 88 official constellations.

corona The high-temperature, outermost atmosphere of the Sun. It is visible from Earth only during a total solar eclipse.

double star Two stars that appear close together in the sky. Optical doubles are chance alignments of the stars; binary or multiple systems are linked by gravity.

eclipse When one celestial body passes in front of another, dimming or obscuring its light.

ecliptic The apparent path of the Sun around the celestial sphere.

electromagnetic spectrum The entire range of radiation: radio waves, infrared, optical light, ultraviolet light, X-rays, and gamma rays.

emission nebula A cloud of gas glowing as the gas re-emits energy absorbed from a nearby hot star.

equator The imaginary line on a celestial body that lies halfway between its two poles.

equinox The moment when the Sun appears to stand directly above a planet's equator.

escape velocity The minimum speed an object (like a rocket) must attain in order to travel into space. If too slow, gravity will pull it back.

galaxy A huge gathering of stars, gas, and dust, bound by gravity and having a mass ranging from 100,000 to 10 trillion times that of the Sun.

gas-giant A large planet composed mainly of hydrogen (Jupiter, Saturn, Uranus, and Neptune).

globular star cluster A spherical cluster, some with over a million stars, most of them old and red.

Kuiper Belt A region beyond Neptune's orbit where multitudes of icy objects orbit the Sun.

light-year The distance that light travels in one year, about 6 trillion miles (9.5 trillion km).

Local Group A gathering of over 30 galaxies, including the Milky Way.

meteor The bright streak of light produced by a piece of space debris as it burns up in Earth's atmosphere.

meteorite A piece of space debris that reaches Earth's surface intact.

nebula A cloud of gas or dust in space; it may be dark or luminous.

neutron star A massive star's collapsed remnant, made up almost completely of very densely packed neutrons. May be visible as a pulsar.

nova A white dwarf star in a binary system that brightens suddenly by several magnitudes as gas pulled away from its partner star explodes in a thermonuclear reaction.

nucleus Core of a galaxy or comet.

Oort Cloud A swarm of billions of comets extending out about 2 light-years from the Sun.

open star cluster A group of a few hundred relatively young stars.

orbit The path of an object as it moves through space under the control of another's gravity.

planetary nebula A shell of gas blown off by a low-mass star when it runs out of fuel in its core.

planetesimal A small, rocky body; one of the small bodies that coalesced to form the planets.

pulsating variable A star that changes its brightness as it expands and shrinks regularly.

red giant A large, cool, red star in a late stage of its life.

reflection nebula A cloud of dust or gas visible because it reflects light from nearby stars.

satellite Any small object orbiting a larger one, such as a rocky or artificial object orbiting a planet.

solar day The time from one noon to the next. Earth's solar day lasts 24 hours.

solstice The time of the year when a planet tilts most directly toward (or away from) the Sun.

sunspot A dark, highly magnetic region on the Sun's surface that is cooler than the surrounding area.

supercluster A cluster of clusters: a vast assemblage of entire clusters of galaxies.

supernova The explosion of a massive star, briefly outshining a galaxy, that occurs when the star reaches the end of its fuel supply.

terrestrial planet A planet with a mainly rocky composition (Mercury, Venus, Earth, and Mars).

variable star Any star whose brightness appears to change.

white dwarf The small, very hot but faint remnant of a star that remains after the red giant stage.

wormholes Theoretical tunnels through hyperspace. They could provide a short cut to a distant place in our Universe or to another universe entirely. They might also permit time travel.

zodiac The name given to 12 constellations that lie along the path of the Sun projected onto the sky (the ecliptic).

Index

Page numbers in *italics* refer
to illustrations

Acknowledgments

PHOTOGRAPHIC CREDITS

Key t=top; l=left; r=right; tl=top left; tc=top center; tr=top right; c=center; cr=center right; b=bottom; bl=bottom left; bcl=bottom center left; bc=bottom center; bcr=bottom center right; br=bottom right

AAO = Anglo-Australian Observatory; AF = Akira Fujii; APL/CBT = Australian Picture Library/Corbis; AS = Andy Steele; BA = Bridgeman Art Library; CIC = Celestial Image Co.; DM = David Miller; DS = Digital Stock; EIT = Extreme Ultraviolet Imaging Telescope consortium, ESO = European Southern Observatory; NASA/EO = NASA/Earth Observatory; NASA/ES = NASA/Earth from Space: NASA/GRIN = NASA/ Great Images in NASA; NASA/HST = NASA/Hubble Space Telescope; NASA/JPL = NASA/Jet Propulsion Laboratory; NASA/SF = NASA/Spaceflight; NASA/VE = NASA/ Visible Earth; NASA = National Aeronautics and Space Administration; NOAO = National Optical Astronomy Observatory; OS = Oliver Strewe; PD = Photodisc; PE = PhotoEssentials; PL = photolibrary.com; SB = Stockbyte; SOHO = Solar and Heliospheric Observatory

12bc NASA/ES br NASA/VE 13c NASA/EO/Professor Stanley Herwitz/Clark University 18bl SOHO r NASA/JPL/SOHO/EIT 20tr NASA 26br NASA 34bl NASA/JPL 44br NASA/ES/Lyndon B. Johnson Space Center/Astromaterials Research & Exploration Science/Earth Sciences & Image Analysis 52bcl, bcr, bl, br DS 53bl DS 53bl George East/PL tr APL/CBT 60bc NASA/JPL 62br NASA 68bl br DS 74l NASA/GRIN 86r NASA/JPL 89bl NASA/JPL/USGS 92bl NASA/HST/Dr. R. Albrecht/ESA/ESO 94bl APL/CBT 98bl PD br DS 99c PD 104c NASA/ESA/J. Hester 109cr NASA/Don Figer 116bl AAO/ROE 118c NASA/HST/C. R. O'Dell et al l NASA/J.P. Harrington/K.J. Borkowski 119l NASA/B. Preston/STScI/Max-Q Digital 120r DS 121c NASA/JPL/Space Telescope Science Institute 124l NASA r NASA/2MASS/Atlas Image 128bl, br NASA/STScI/AURA c NASA/AURA/STScI/J. Gallagher et al 130bl NASA/STScI/AURA 131tr NASA/KPNO/NOAO/Roger Lynds et al. tl NASA/AURA/STScI 150bl APL/CBT r PL 151br NASA/JPL/Cornell tc NASA/ES 154l NASA r NASA/GRIN 155c NASA/GRIN 156r PD 157r Johnson Space Center Office of Earth Sciences 159r DS 160l PD r NASA 161c DS 162bl NASA/Human Space Flight 164c, r DS 165l PE 168bl SB 172cr NASA/SF/Nasa Dryden Flight Research Centre Photo Collection l NASA/SF 174br PL/NASA 175c DS 184bl APL/CBT br PD 185bl BA 186br APL/CBT 189l Northwind Picture Archives 199tr PL 202r NASA 206c APL/CBT 208bl Pekka Parviainen/PL br AF 209l AF 214b OS 215r AF 221r AF 257c AF 258bc DM bl, br PE 259c NASA/APOD/C. Mayhew/R. Simmon/GSFC/NOAA/NGDC/DMSP Digital Archive 260bl, r OS 262bc, bl Meade Instruments Corporation 263br OS 268bl OS 269c OS 270bl OS br PD 271c DS 272bl OS 273c AF 274c PE 275tc, tl DS tr PD 276br DS 277c DS 278bl NASA/JPL 279c DS 280l John Thomas/PL r Jack Finck/PL 281c Eurelios/PL 282bc, bl AF 283l, tr AF 284bl AF 285bl, tl AF tr Rev Ronald Royer/PL 286bl AF 287c PL 288bl AS 289c AS 290r NOAO 291r NASA/APOD 292bl NASA/STScI/AURA 293c DS 294bl DS 295c CIC/PL 296br NASA 297l PL/NOAO 298br CIC/PL 299c PD 300br PL/MPIA-HD/Birkle/Slawik 301c PL/ESO 302bl PL/NOAO 303c PL/CFHT/Jean-Charles Cuillandre 304bl Dr Luke Dodd/PL 305c ESO

ILLUSTRATION CREDITS

Gregory Bridges, Lynnette R. Cook, Andrew Davies/Creative Communication, Chris Forsey, Mark A. Garlick/www.space-art.co.uk, The Granger Collection, Rob Mancini, Moonrunner Design Ltd., Trevor Ruth, SOHO, Marco Sparaciari, Steve Trevaskis, Thomas Trojer, Rod Westblade, Wildlife Art Ltd. (Julian Baum/Tom Connell/Luigi Gallante/Lee Gibbons/David A Hardy/Sandra Pond), Simon Williams/Illustration, David Wood

Star Maps: Wil Tirion

CONSULTANT

Dr. John O'Byrne is a Senior Lecturer in Physics at the University of Sydney, NSW, Australia, and is Secretary of the Astronomical Society of Australia.